Curing

of

Fishery Products

Teaparty Books

Curing of Fishery Products

by Norman D. Jarvis, Technologist

Curing of Fishery Products was originally published in 1950 by the Fish and Wildlife Service of the United States Department of Agriculture.

Library of Congress Catalog Card No. 87-50413
I.S.B.N. 0-9610602-5-5

First Edition
1 2 3 4 5 6

Printed in the United States of America

Curing of Fishery Products

Contents

Curing of Fishery Products

Introduction

Fish curing comprises all methods of preservation except refrigeration and canning. It includes (1) the drying, smoking, salting and pickling of fish, (2) various combinations of these methods, and (3) miscellaneous methods such as the use of vinegar and fermentation processes or ripening.

The drying and smoking of fish are ancient processes. Archaeologists and anthropologists tell us that drying and smoking were probably developed shortly after the discovery of fire and before man learned to make pictographs on rocks. The art of salting is also very old, going back to the Stone Age. The use of vinegar and spices goes back, at least, to the Greeks and Romans.

Fish curing has been an important factor in the development of all great maritime nations. From the Middle Ages until well into the nineteenth century agricultural methods were poor. Livestock were wintered with difficulty, and meat curing methods were so poor that there was little meat during the winter, also, other protein foods were scarce. Canning and freezing had not yet been developed. The people needed a fairly low-priced protein food which would remain in good condition for some time. Dried, salted, and smoked fish supplied this need. Before the beginning of the industrial age, international commerce was largely in natural products. The wine of Spain was exchanged for the dried and salt fish of England, the Netherlands, and Norway. The smoked herring of England was traded throughout central Europe. Even in the sixteenth century, when England was fighting Spain, her trade in dry fish with Spain was continued. It has been reported that England gained more wealth from cured-fish products in the sixteenth century than Spain obtained in gold from the Americas.

Despite the importance of fish curing since prehistoric times, information on the methods of the industry is scattered and exact data are lacking. Great interest is exhibited in fish curing today, as it does not require the equipment or capital needed for canning or freezing. An individual with a little capital may dry, salt, or smoke fish, using only home-made equipment. Species of fish may be utilized that are not suitable for canning. Localities not adaptable to the operation of canning or freezing plants may be used for fish curing. During the years to come the whole world will need all available sources of protein. Dried and salt fish are not only good sources of protein but may be produced at lower cost than other animal protein foods. Statistical data given elsewhere in this report show that fish curing makes up the greatest volume of fishery products in every country in the world having important fisheries, excepting the United States and Canada. In the latter country, cured and canned products are approximately equal in amount.

Technical studies of the principles on which fish curing is based, and on improvements in methods and equipment together with descriptions of standard methods have been made from time to time. Some of the publications are unavailable for reference, and the work has not been integrated as a whole. The Fish and Wildlife Service receives a large number of requests from individuals, commercial firms, and official agencies, both here and abroad for up-to-date information on the technology of fish curing. This situation emphasizes the need for a reference handbook on commercial curing of fishery products. This report is intended to supply that need.

Chapter One

Fish Curing in the History of the United States

ATLANTIC COAST

The fish-curing industry of the North Atlantic coast of North America dates back at least to the year 1500. There are authentic records of fish-curing activities as of that period, and legends of activities at much earlier date. An extensive fish-curing industry along the North Atlantic coast of North America was carried on for more than one hundred years before there was any permanent settlement. As early as 1580 more than 300 ships from Europe were salting cod in this area. Newfoundland, "the oldest British colony," owes its origin to the fish-curing industry which is still the dominant factor in the economic life of that country.

The early colonists in New England and the Maritime Provinces would not have been able to survive without the salt cod and smoked herring they could prepare, for soil was poor and the climate uncertain. While fish meant food to the early colonists, cured fish soon became their capital resource and their stock in trade for the purchase of supplies. Their most abundant fish, cod, could be manufactured into a durable protein food product, withstanding the primitive shipping and storage conditions of the day, and was comparatively low in price. Other cured fish such as smoked halibut and herring, pickled sturgeon, and salt salmon were soon being shipped abroad. Out of this grew the "triangular trade": salt fish to Europe, manufactured goods from Europe to the West Indies, and sugar, rum, and molasses to New England. The trade in salt fish stimulated other industries and capital was gradually accumulated so that the colonists could go into shipping. Later, other natural resources such as timber were exploited, and the first attempts were made to create other local manufactures.

The importance of fish curing in the development of trade and industry in New England is discussed at some length by Ackerman (1941) who describes how various New England industries owe their origin to the foreign trade in cured fish.

The fish-curing industry continued to grow and prosper, dominating the economic life of the New England colonies in the late 17th and 18th centuries. The French government was also concerned. It was attempting to expand the fisheries in the northwestern Atlantic in order to build up a large fish-curing industry in its colony in Canada. Both the British and the French attempted to dominate as much of the North American fishing grounds as possible, in order to secure the trade in salted and dried fish for themselves. Attempts were made to establish boundaries, but they were poorly defined, so that the fishing rights over a large area were the cause of frequent bickering, sometimes flaring up into undeclared warfare. Indeed, the series of wars between France and England were in part due to disputes over fishing grounds and fish-curing locations. The fishermen and fish curers of New England and Nova Scotia played an important part in England's conquest of Canada, for to them the fishing rights meant life or death.

The disputes did not end with the ousting of France from Canada, but continued with the New England colonists, on the one hand, and the English, on the other. Parliament passed a bill in 1775 which prohibited the New England colonies from trading directly with foreign countries and prevented New England vessels from fishing on the banks of Newfoundland, in the Gulf of St. Lawrence, and on the coasts of Labrador and Nova Scotia where they had been accustomed to go. This meant ruin to the New England fish-curing industry, and the edict was one of the leading causes of bringing the New England colonies into the Revolutionary War.

The treaty of peace negotiated in 1783 was delayed by the insistence of the American delegates on securing favorable fishery rights. They regarded these so important that they refused to sign a general treaty of peace, leaving the fishing rights for later adjudication. The British proposed to exclude the New England fish-curing industry from grounds and areas where New Englanders had fished and cured fish extensively as colonists. Finally, however, the American delegation was able to obtain a treaty article on fisheries which granted favorable conditions to the United States.

The New England fish-curing industry generally prospered under the new Republic, taking salt-cod markets in southern Europe and the Mediterranean from the British and Scandinavians. Disputes arose with Great Britain over trade, the interpretation of fishery rights, and the impressment of American fishermen and seamen into the Royal Navy. Restriction and embargoes were imposed by both Great Britain and the United States, resulting in a decline in the salt-fish industry after 1807. The War of 1812 almost ruined the industry. The war was so unpopular among shipping, commercial, and fish-curing groups that there was a move toward secession in some of the New England states.

At the end of the War of 1812 the British claimed that the war abrogated the treaty of 1783. The United States claimed that this treaty was still valid. Seizures of American vessels were made and it seemed for a time that a new war might break out. Tension was eased by the signing of a new fishery convention in 1818. It was followed, however, by a whole series of disputes on interpretation, at times resulting in severe diplomatic tension. The rights involved in our fish-curing industry in the northeastern Atlantic off the coasts of Canada and Newfoundland were probably the most important single cause of disagreement during most of the nineteenth century between the United States on the one hand, and Great Britain, with the Dominion of Canada, on the other.

Trouble occurred more frequently and was of less intensity in the last decades of the 19th century, as refrigeration developed and a wider market was created in the United States for fresh fish, making salting and drying of fish on the northeastern coast less important. It was also possible to conduct all operations at sea away from the coast. While the feeling was not so intense, the points of difference in interpretation of the Convention of 1818 remained unsettled and it was decided to submit the points at issue to arbitration under the Hague Convention. A decision was rendered in 1910. Since that time the only fisheries disputes between the United States, Great Britain, Newfoundland, and the Dominion of Canada, have concerned the duties to be levied on fishery products under our tariff acts. The U.S. Tariff Commission in Report No. 152 gives a complete account of the fishery question on the northeastern Atlantic coast, together with the various treaties and other agreements on the fisheries.

PACIFIC COAST

The fisheries on the Pacific coast of North America have not affected our international relations to the same extent as the fisheries of the Atlantic. This is (1) because their development is much more recent and (2) the development has been different in character. The second factor is possibly due to the fact that development occurred at a period when canning and refrigeration were replacing curing as the principal methods of preservation. Then, too, more of the fishing takes place in clearly defined territorial waters.

There have been a number of disputes between Canada and the United States over Pacific fishery problems but they have been minor in character when compared with those in the Atlantic, and none has concerned fish curing. In the 1930's, Japan invaded fishing grounds off the coast of Alaska, and interfered with our vessels which were catching and salting cod. The cod fishermen threatened to shoot any Japanese obstructing their operations. Japanese fishing was a matter of great concern to the Pacific coast fisherman, but little notice was taken nationally until it was proved that the Japanese were catching salmon despite an undertaking to the contrary. The controversy was still unsettled when Japan went to war with the United States. The fishery dispute may be regarded as a contributory, though minor, cause leading to war with Japan.

Cured fish of various types were the first manufactured products prepared on the Pacific coast. The Indians had a considerable dried salmon industry at The Dalles, on the Columbia River, long before the coming of the white man. The fish were traded to the plains tribes of the interior. The Indians still dry small amounts of salmon at The Dalles for their own use. According to Cobb(1930) the Russians operated a commercial salt-salmon industry in Alaska at the beginning of the 19th century. Salt salmon was shipped as far as St. Petersburg. Soon afterward the Northwest Fur Company started a salmon-salting business on the Columbia River. The Northwest Company merged with the Hudson's Bay Company which shipped salt salmon to Hawaii, Australia, China, Japan, and the eastern United States. American fishermen salted salmon in Alaska while it was still a Russian possession. A number of large salmon canneries of today were originally established as salmon salteries.

The presence of cod off the coast of Alaska was established in the 1860's and the possibility of building a prosperous salt-cod industry there was one of the arguments advanced in congress for the purchase of Alaska. More recent but still incomplete studies have established that the Pacific banks are larger and of greater potential production than the Grand Banks off the coast of Newfoundland. Yet, utilization today is less than it was 30 years ago.

During World War II the Pacific coast fish-curing industry was much more adversely affected than that on the Atlantic coast. When the Alaskan area became a combat zone, almost all fishing and fish-curing activities were stopped. All but one of the cod-salting vessels was requisitioned by the Government. Herring salting in southeastern Alaska was almost impossible because of the great increase in costs, the scarcity of labor, and the necessity of using commercial vessels for war purposes. These causes also adversely affected the mild curing of salmon. The loss of foreign markets and the effect of pricing regulations were other unfavorable factors. It is expected that there will be some recovery of the Pacific coast fish-curing industry in the postwar period, but anything approaching a restoration of the pre-war trade will require the development of specialty products and a reduction in the cost of production.

Chapter Two

Statistical Review of the Fish Curing Industry

Complete statistical data on the production of cured fish in the United States over the past 60 years, or any consecutive series of years within the period, are lacking. This is unfortunate as both canning and refrigeration industries were being developed during this period; in fact many changes were taking place in food habits and food products, with great effect on the fishing industry.

The first production figures available on the fish-curing industry are reported by Goode (1884), but the data do not cover the entire United States and are reasonably complete only for the New England states. The first complete figures on manufactured fishery products are for 1908. It is the first reasonably complete and accurate statistical record on the production of cured fish in the United States. In 1908 the total production amounted to 187,299,000 pounds with a value to the producer of $11,018,000. In 1940 production decreased to 97,439,000 pounds but the value was increased to $14,235,000. The decline in the production of salt groundfish is even more startling if the New England

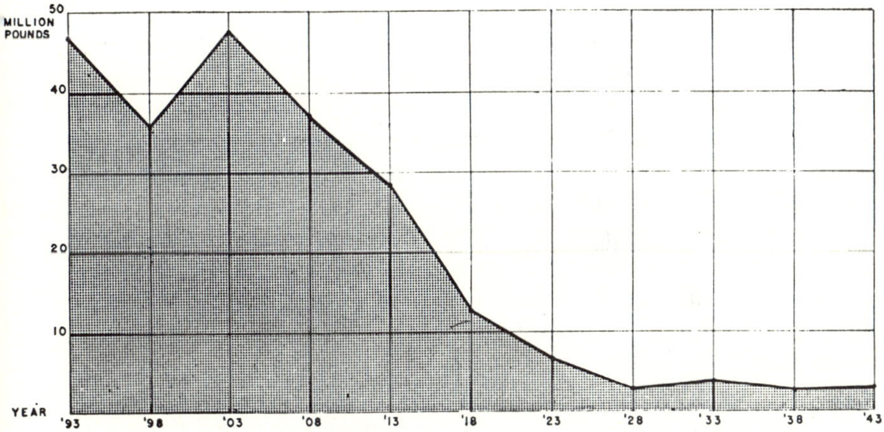

Figure 1.—Landings of salt groundfish, New England, 1893-1943.

area, historically the most important, is considered separately. This is shown graphically in figure 1. From 79,009,000 pounds in 1880 the production fell to 4,742,000 pounds in 1919.

In 1908, dried and salted groundfish cod, haddock, cusk, hake, and pollock) totaled 84,642,000 pounds while in 1940 it was 17,697,000 pounds. It is believed that a portion of the boneless salt cod reported in 1940 was prepared from green-salted fish imported from Canada and is, therefore, not entirely a product of the United States fisheries. Pickled or hard-salt salmon amounted to 14,595,000 pounds in 1908 while in 1940 it was 829,000 pounds. The production of mild-cured salmon showed little change since it equaled 8,483,000 pounds in 1908 and 7,770,000 pounds in 1940. Incidentally, 1940 was a war year, and foreign markets which usually absorbed an important share of the production had been eliminated. More than 1,200,000 pounds of halibut were smoked in 1908, while only a few hundred pounds were prepared in 1940. Production of smoked herring was 13,311,000 pounds in 1908 but it totaled only 3,629,000 pounds in 1940. On the other hand, no smoked kippered sablefish was prepared in 1908 while there was a production of 1,176,000 pounds in 1940. The production of dried shrimp is given as 342,000 pounds in 1908 and 2,069,000 pounds in 1940. Herring pickled in vinegar and spiced herring were not reported in 1908, while production of these two items totaled 3,537,000 pounds in 1940. This is believed to involve some duplication as herring pickled in vinegar is used in the production of spiced herring. Kippered salmon is not reported in 1908 while 2,543,000 pounds were produced in 1940. Production of smoked salmon amounted to 4,297,000 pounds in 1908 and it was 8,423,000 pounds in 1940. No smoked buffalofish, butterfish, or carp are listed in 1908, while production of these three items totaled 1,004,000 pounds in 1940.

In 1908 cured fishery products represented 42 percent of the manufactured fishery products while canned fish and sea foods amounted to 57 percent (Census 1911). In 1940 production of cured fishery products amounted to 97,439,000 pounds with a value to the primary producer of $14,235,000 while production of canned products amounted to 708,930,000 pounds with a value of $94,182,000. Thus in 1940, cured fish accounted for 9 percent of the quantity and 10 percent of the value.

The fish-curing industry is much more important in other countries with large fishery industries, than in the United States, and fish curing in these countries does not show a decrease to the same extent as in the United States. Japan was the world's largest producer of cured fishery products at the outbreak of World War II. In 1936, the most recent year for which statistics are available, Japanese data show a production of 1,234,709,000 pounds of cured fishery products with a value of $49,609,000. The production of canned fish amounted to 223,899,000 pounds, with a value of $20,318,000. It is understood that cured products were prepared largely for home consumption, and the greater portion of the canned products was exported. Fish paste was the most important single item, with a production of 203,856,000 pounds. Seaweed prepared by several methods is also an important cured

Figure 2.—Air drying salmon, Alaska.

product in Japan. The only method of curing not greatly favored in Japan is smoking. The total amount cured by this method was only 1,847,000 pounds.

The production of cured fishery products in Canada is greater than in the United States but the value to the producer is less. The total of cured products in 1938, the last year of normal prewar production, amounted to 125,451,000 pounds with a value of $4,250,000. Difference in value of production is believed due to the fact that the items produced in Canada are largely the standard low-priced products, while the United States prepares more high-priced cured fish specialties. For instance, in Canada the total production of smoked and kippered salmon in 1938 is reported as 26,000 pounds with a value of $3,000, while the United States production of smoked and kippered salmon in 1940 was 10,966,000 pounds with a value of $3,563,000. Canadian statistics do not show any production of such articles as smoked sturgeon and spiced herring.

Causes of Decline in Fish Curing in the United States

The decline in production of cured fishery products cannot be ascribed to any one cause. An important factor is the development of better methods of preservation, such as canning and freezing. Not only has the consuming public come to prefer canned and frozen fishery products rather than cured, but the industries based on these products took material formerly utilized in fish curing. As an example, haddock was in little use in the fresh-fish market of the 19th century but appeared almost entirely as a dried-salted or smoked product. The New England states prepared 6,510,000 pounds of boneless and dry-salted haddock in 1908, but only 13,000 pounds in 1940. No fresh or frozen haddock fillets are reported in 1908 but 36,597,000 pounds are reported for 1940. The development of the fillet industry practically eliminated the curing of haddock, but the total consumption of haddock has greatly increased.

The new methods, canning and freezing, resulted in a more marketable product and in the case of canning at least, a greater length of preservation. Cured fish was the most durable preserved protein food available until the latter part of the 19th century, but admittedly when prepared for the table it does not resemble fresh fish as closely as canned fish does. It is subject to deterioration by reddening, molds, and other types of spoilage to which canned products are immune, and does not offer the variety available in canned fishery products.

Cured fish, as prepared in large-scale production, is not so economically practicable as canned or frozen packaged products. The percentage of loss in preparation is much greater so that raw material costs per unit of consumption are higher for cured products. The curer, therefore, cannot afford to pay as much for raw material. Fish curing to date involves a great deal of hand labor and except to a limited extent, in Norway and Germany, has not been mechanized while canning has been highly mechanized. Fish curing is, therefore, unprofitable when labor and raw material costs are high. Cost of production adversely affected the United States fish-curing industry earlier and to a greater extent than that in Newfoundland, Iceland, Norway, and Canada. This country was forced to abandon foreign trade in cured groundfish after the first World War, but the decline was evident even before this. Later, even the domestic trade became unprofitable.

The most pronounced trend in the food industry has been toward the consumer-size package, and there has been much improvement in packaging. Fish-curing methods are essentially those of 60 years ago. Much of the cured fish is handled in pails, barrels, or in bulk. Therefore, most cured fish are not attractive for the retailer to handle.

Another factor in the decline of fish curing is the change in food habits. European immigrants were accustomed to cured fishery products and introduced many different cured-fish articles to this country, but their children developed other food habits.

Prospects for Increasing Consumption

Curing cannot be expected to regain its place as the principal method for the preservation of fish and sea foods, but the situation is not so dark as the preceding discussion would seem to indicate. The trade in cured fishery products can be increased. In the first place, there must be a standardization of quality. There is too wide a variation in quality between the products of individual producers and day-to-day production. Improvement requires careful study of methods used and the development of controlled techniques.

The packaging of cured fishery products should be given the same careful study as the packaging of fresh and frozen foods. Some progress had been made prior to World War II by individual firms. While this work was limited in scope, and results are not sufficiently clear to warrant positive conclusions, it is indicated that proper packaging is a tremendous aid in selling and would make cured fish more attractive to the retail dealer and consumer.

A well prepared and consistent educational program carried on over a period of years, spreading knowledge of the principles of food preservation and advocating techniques developed through systematic experimental work, would acquaint the general public with the superior appetite appeal and nutritive value of well-prepared, cured fishery products. This program should include demonstrations on cookery and preparation for the table, which has almost become a lost art. The industry is now realizing that spreading information of fresh-fish cookery has been neglected, with a consequent effect on demand. This is even more true with cured fish.

It is also believed that a variety of specialty items rather than a few old-time standard articles would increase sales. These items can generally be sold at higher prices so that labor and raw material costs are not a production obstacle. An example of this is the increased production and demand for smoked salmon. The demand for various spiced-herring specialties is also increasing. Smoked oysters, shrimp, and crab meat are new cured-fish specialties in excellent demand. Many specialties are described in this report and from the information given here it should be possible to develop others.

Chapter Three

Preservative Action in Fish Curing

Food preservation attempts to prevent or inhibit deterioration or spoilage of the food material for as long a period of time as possible. The agents causing spoilage may be present in the raw material when it is obtained; they may be introduced later during the process of manufacture or preservation, or during the storage period. There are two general types of deterioration: (1) that due to microbiological action, and (2) changes as a result of chemical or physical action. Micro-organisms are responsible for by far the greater amount of food spoilage.

SPOILAGE ORGANISMS

The number of types of spoilage organisms which may be involved in the deterioration of cured fishery products is almost infinite. They may be divided into three general classes: (1) molds, (2) yeasts, and (3) bacteria. In canning products, heat- resistant spore-forming bacteria are almost entirely responsible for any spoilage which occurs. The ordinary heat treatment given in canning is sufficient to render impotent all yeasts, molds, and ordinary types of bacteria.

Molds

Molds are fungi which are distinguished from micro-organisms by the formation of mycelia, which are networks of filaments or threads. These threads or hyphae are usually visible to the naked eye. Molds as a group are mostly aerobic, preferring a slightly a slightly acid medium, and can tolerate higher acidity and osmotic pressure than bacteria. Molds in general require less moisture than bacteria. They are essentially surface growers, though the hyphae will soon penetrate throughout the whole flesh structure. The mold spores are much better adapted to air dispersal than bacteria, and, therefore, there is more danger of contamination of smoked or dried fish. The two groups of molds most likely to cause trouble in cured fishery products are species of *Penicillium* and *Aspergillus.* For more adequate information the reader should consult Tanner (1944) or any other good text on microbiology.

Yeasts

Yeasts are characterized as fungi which multiply by budding or fission or a combination of these two processes. They may also reproduce by intracellular spore formation whereby each cell may develop into an ascus containing several ascospores. Yeasts are generally aerobic. They usually require some sugar. Yeasts, while of great importance in certain food industries are of minor importance in fish curing. They may be found in some types of spiced-fish products, and the preservation of certain cured fishery products in the Orient such as fish in rice paste is due to fermentation by yeast.

Bacteria

Bacteria are unicellular organisms which reproduce by simple binary fission. Bacterial cells may be divided into three basic groups or shapes: the cocci or spheres, the bacilli or straight rods, and the spirilla or curved rod forms. According to Cruess (1938), the grouping should include practically all the fission fungi. The deterioration or spoilage of fish flesh is particularly due to bacteria. Fish flesh is especially susceptable to both bacterial and enzymatic decomposition, so that the problems of preservation are particularly acute. Bacteriologists have given considerable attention to the microbiology of fish. Tanner (1944) gives a summary of their findings.

Optimum Conditions for Micro-organisms

Micro-organisms require moisture and a favorable temperature for development because they can utilize only liquid food which is absorbed through the cell wall. A moisture content of less than 35 percent has a direct inhibiting effect. The most favorable temperature for the development of ordinary forms of bacteria is between 70° and 100° F. Microbial activity is greatly reduced at temperatures around 32° F. and growth of bacteria is practically inhibited at 0° F. Most organisms grow best in an alkaline or neutral medium so acids have an inhibiting effect. The addition of acids to inhibit bacteria is the basis of the preparation of vinegar-cured fish. Preservation in this case is temporary, however, since a vinegar solution in excess of 5 percent acidity (as acetic acid) will make the product unpalatable. A 15 percent solution is required to absolutely inhibit bacterial activity.

The food must be liquified so that the micro-organisms can absorb the nutrients necessary for growth. The micro-organisms present secrete an exoenzyme which breaks down the substrate into products that can be absorbed through their body walls. Waste products resulting from this absorption are excreted through the cell walls, thus furnishing such distinguishing characteristics of decomposition as "off" odors and tissue break-down. Putrefaction results from the anaerobic decomposition of the flesh.

Autolysis

The cells in flesh tissue also contain enzymes which do not cease their activity upon the death of the animal, and may digest the cells in which they are contained. This process is known as "autolysis" of self-digestion and is distinct from putrefaction as described in the preceding paragraph. Autolysis is especially marked in species which have been feeding when caught, then death prevents completion of the normal cycle of digestion. The larger quantities of digestive enzymes manufactured during the feeding processes are prevented from fulfilling their function of preparing food for absorption. They therefore attack the walls of the digestive tract which are destroyed with extreme rapidity. The enzymes then pass to the flesh which is also softened in a short time. This process is normal and the flesh is not absolutely inedible but it is unsuitable as raw material for curing due to the extreme softness of texture and the "off" flavors developed.

A good example for autolysis, for the purpose of illustration, is that in "feedy" salmon. Troll-caught salmon must be eviscerated immediately after catching and packed in crushed ice. If this is not done they will be found unfit for use in a very few hours. The flesh becomes soft and flabby, the abdominal surface to such a degree that the bones protrude. Though the flesh is not putrefied it may show many breaks and it is difficult to handle in mild curing. The cured product is objectionable in appearance, color, flavor, odor, and texture. Deterioration of Atlantic mackerel occurs even more rapidly than salmon.

Chapter Four

Air-Dried Fishery Products

PRINCIPLES OF FISH DRYING

The two factors necessary for bacterial growth are heat and moisture. Of these, moisture is the more important, for if there is sufficient bacterial growth can occur at temperatures near freezing, while if the moisture content is reduced to a low point, bacterial growth is inhibited for long periods. There are several methods used in the fish-curing industry for the removal of moisture such as exposure to the open air, pressure, use of absorption pads, application of the principle of osmosis as in covering the fish with salt, the use of heat, and combinations of these methods. In curing cod, part of the moisture is removed by salt, part by pressing, and part by exposure to the open air.

The only method of fish curing that can be considered as drying in the strict sense is natural air drying or in mechanical dryers. Removal of moisture by salt is through osmosis, smoke acts as a chemical preservative as well as a drying agent, and pressure forces out moisture. Dehydration is differentiated from drying in that the product is much less like fresh fish than the naturally dried product, moisture content is reduced to a lower point, the form is changed, and the method depends on the successful mechanical control of heat and moisture.

Natural air drying is the oldest method of food preservation. It was used by the ancient Egyptians, and, it is believed, long before them by the most primitive tribes of the early stone age. Air drying requires either a warm climate with low humidity, or, if the climate is cold, a good breeze and low humidity. It is used extensively commercially in Norway and in the Orient, especially in India, China, Japan, and the Philippines. While the climate of the United States is not particularly suitable for air-drying, it is used for curing a variety of products in this country. Of these, shrimp is the most important commercial commodity. A large amount of salmon is cured by air drying in Alaska, principally as winter food for the natives and their dogs.

Halibut Rackling

This is a product prepared largely on the north Pacific coast of the United States and Alaska. It was introduced in the United States by Scandinavian fishermen, who prepare it mostly for home use. It is a well-known commercial product in Norway, where it is a prominent product of the Bergen fish market.

Rackling is prepared principally from halibut on the Pacific coast, but large flounder or rock cod, or similar fish of fat content of about two percent are suitable. The fish are headed and cleaned, leaving the collarbone or napes. After removing the viscera the fish are split into two sides, removing the backbone. The sides are cut in long narrow strips about an inch wide which are left joined together at the collarbone. The pieces are washed very thoroughly. All traces of blood must be removed or the product will be inferior when dried.

After draining, the strips are soaked in 95° salinometer salt brine (one that will float a potato) for one or two hours. The fish are then hung out to dry where they will be exposed to as much breeze as possible, preferably in a shady place where they will not be exposed to direct sunlight. Often, however, the strips of fish are hung out to dry in the rigging of fishing vessels.

Drying requires from one to two weeks. Rackling is most often eaten like jerked meat without any preliminary preparation. It may be prepared for use by soaking for a few hours to rehydrate, then it may be steamed or boiled, and made into various dishes such as fish cakes, fish loaf, or creamed fish.

Salmon

The air drying of salmon is an important method of preservation in Alaska. It is used mostly by Indians. Drying is carried on most extensively in the interior, especially along the Yukon, Kuskokwim, Tanana and Copper Rivers, but it is also followed all along the coast of Alaska. Dried salmon is a staple winter food for the natives of Alaska, especially for their dogs. A sledge dog is reported to require 10 pounds of dried salmon per day in winter when in harness. Whitefish and grayling are also dried by the same method in the interior, and cod, halibut, and large flounder along the coast. Statistical data indicate a production of 2,000,000 pounds of dried salmon in 1940 (Fiedler 1943). Since much of the production is at scattered and isolated spots, the amount may have been larger.

The salmon taken for drying are caught principally by fish wheels along interior streams. The apparatus and method are described in detail by Cobb (1930). Along the coast, stake gill nets are used extensively. Some salmon are taken for drying by haul seines, both on the coast and in the interior. In rivers of the Bering Sea area, salmon are caught for drying by means of wickerwork traps constructed on the principle of the fyke net. These traps are made of a series of cylindrical and conical baskets, fitting into each other with a small opening in the end connecting one with the other, the series ending in a tube with a door in the bottom through which the catch is removed. The work of preparing dried salmon is performed mostly by women.

Good fish curers break the backbone just back of the head as soon as the salmon are caught. This is done to bleed the fish, and also to kill them at once, preventing threshing about and bruising the flesh. The butchering or cleaning process is as follows: The heads are cut or broken off, leaving the collarbone or nape. A knife is inserted at the collarbone just above the backbone and is drawn along the backbone to within two or three inches of the tail. A similar cut is made just under the backbone, which is then broken off close to the tail. The salmon are now divided into two halves except for the small section near the tail where they remain joined. The viscera, membranes, and other offal are scraped out, and the fish are washed roughly, in sea water, if along the coast. A series of transverse cuts, about three or four inches apart, is made to facilitate drying.

The fish are then hung from poles on a drying frame (fig.2). They are hung flesh side out, with one side on each side of the pole. Sometimes when poles cannot be obtained for a drying rack, the two sides of the fish are separated and laid flesh side up on a gravel beach. The salmon do not dry so well when this is done.

The salmon remain on the rack until thoroughly dried. Sometimes in rainy or foggy weather the fish are taken down and placed under shelter, but as a rule they are left on the racks with the skin side turned outward. When the fish are of average size the drying process requires from 10 days to 2 weeks in ordinary weather. More time is required if drying conditions are not good, or if the salmon are large. The location of the drying racks is important. They should be placed on a headland or in an open space on dry ground where there is a good sweep of breeze. On still days, a smudge of green wood is sometimes use to keep the flies away.

When the drying process is completed, the dried salmon are stored in a cool, dry place, out of the reach of children and dogs. The storehouse is often built on a platform elevated 10 to 12 feet from the ground and is the so-called cache of Alaskan stories.

Sharks, Skates, and Rays

Shark meat is dried for food in numerous parts of the world where these fish are taken in quantity. Various species of the shark family are probably the most abundant source of raw material for fish curing on the Pacific coast of South America. These fish are dried commercially on the northern section of the coast of Peru. Skates and rays are also dried on a fairly extensive scale. With considerable development taking place in the fisheries of Peru, it is expected that production will be increased.

For a tropical area the Pacific coast of South America has unusually favorable conditions for fish drying. This is because the climate is very dry, the temperature moderate, and there is a regular breeze in many sections. The principal handicaps to date have been lack of fishing craft, poor transportation facilities, and unorganized marketing. The information given here was gathered on a survey made during 1941, at the request of the Peruvian Government (Fiedler, Jarvis, and Lobell 1943).

Sharks and rays (*Elasmobranchii*) are dried, such as "angelotes" (angel shark, *Squatina squatina*), "tollos" (dogfish or small sharks), "cazons" (large sharks of several varieties, none of which were identified as to species), and "rayas" (several species of Rajidae).

When the fish have been landed on the beach they are divided into individual shares and each fisherman sets to work on his own pile of fish with a home-made sheath knife. In butchering angelotes, tollos, and cazons, the fins are cut away, then a first cut is made, splitting the fish;

a second cut opens up the right side; while two or three cuts more open up the left side. The head is then cut away, viscera removed, and the flesh is scored, for rapidity in curing. Butchering is done very rapidly and skillfully. The fish are buried in the sand as they are butchered.

The dressed fish are taken down into the salt water, usually by the fishermen's children, who wash the fish and free them from blood, bits of viscera, sand, or other waste material. Two or three thin pieces of cane are inserted into the flesh crosswise, to hold the fish open, and they are hung on pole racks and left to dry in the open air. Drying requires about one week, although it may be finished in four or five days if the fish are small or weather conditions are unusually favorable. No salt or other curing agent is used, preservation depending entirely on air-drying. The cured fish are known as "bacalao" and resemble stockfish in general appearance but are somewhat thinner.

Skates or rays are prepared as follows: After laying the fish on its back, two circular cuts are made down the ventral side. The first slices away the lower wall of the mouth and gill cavity, leaving this wall hanging as a flap on the left side anteriorly. The second cuts away the lower wall of the abdominal cavity, leaving this as a flap on the left side posteriorly. The entrails are then removed; a vertical cut from above is made through the backbone from the head to the base of the tail and one or two short slashes may be made on each side of the thick base of the tail. A series of cuts are then made across the disk of the ray, which now looks like a circular gridiron. These slashes penetrate to the skin below.

After the rays are all butchered, sand is rubbed well into each cut and the fish are laid in a hole in the beach. The hole is covered by straw mats or by a layer of sand. The moist sand absorbs the blood, and the fish are removed after about 24 hours. They are then washed in the surf, drained, and a small amount of coarse salt (half ground) is rubbed into the flesh, after which they are hung across pole racks to dry in the open air. Four or five days are required for drying. This dried fish is said to remain in good condition from two to six months, depending on size, care taken in drying, and storage conditions while held for sale.

Shark Fins

Shark fins are used extensively for food in the Orient, particularly by the Chinese. They are cured in the Philippine Islands, China and other Asiatic lands, the islands of the South Pacific, the West Indies, South

America, Southern California, and Mexico. The United States production of dried shark fins is given as 24,000 pounds in 1940.

While the soupfin shark is taken extensively, it is used principally as a source of vitamin oils. The fins are only a byproduct and are sometimes wasted in this fishery. Any variety of shark will supply suitable raw material, provided the fins are large enough. They should be at least six inches in length, otherwise they are too small to have any substance. The best sizes are from 8 to 12 or 14 inches. The dorsal fin is preferred for drying, but the pectoral and anal fins are also used. The caudal fin is considered worthless.

The fins must be properly trimmed and cured to make an acceptable product. They are cut off at the joint connecting the fin with the body. All fleshy parts must be trimmed away, leaving only the true fin with its rays, as the cartilaginous material of the rays is the only part valued as food. After trimming and washing, the rays are spread on low bamboo or wickerwork frames, or flakes, to dry in the sun. The fins are turned from time to time. No salt or other preservative is used, though some of the curers dust the cut surface of the fin with lime. From two to three weeks are required for drying. The fins are usually brought into a dry shelter at night.

When dry they are packed in cases or barrels of about 250 pounds each or in gunny sacks holding from 150 to 200 pounds. The average wholesale market price is 65 to 75 cents a pound for choice fins. In normal times about five tons per month are handled at San Francisco, where they are bought by the Chinese merchants for export to China (nine-tenths of the total trade). Dried shark fin is called "yu-chee" by the Chinese of San Francisco. It is soaked in water for two or three days, then cooked in soups or broths.

In China there is a more or less elaborate system of grading shark fins. They are classified either as white or black although none of the fins is entirely black or perfectly white. They are divided into a number of groups with different values, depending on color, size, and variety. The most important commercial grades of fins are the following: white spotted fin (boon leong sit), divided into large and small white fins (chu sit); small white fins (peh sit and khiam sit); large black fins (tua sit); small black fins (oh sit and seow oh sit) and small black-tipped fins (oh ku sit) (Seale 1914).

Stockfish

Stockfish of "torfisk," as it is known in Norway, is one of the most important fishery products of that country. Stockfish are cured entirely by air drying without the use of any salt. This product has been a staple article of trade in Europe for at least a thousand years. In the Middle Ages it was one of the few protein foods which could be preserved over a long period. Dry salt fish were not manufactured so extensively then because of the higher cost of salt. Also, conditions of handling in that period required a product with a lower moisture content than the average dry-salt fish.

Figure 3. – Drying Stockfish, Norway.

Stockfish has a large market both in Europe and in Africa. It is also sold in the United States especially in centers with large populations of Scandinavian or Italian origin. All species of the cod family are used in the preparation of stockfish: cod (*Gadus callarias*), haddock (*Melanogrammus aeglefinus*), hake (*Merluccius* sp.), cusk (*Brosme brosme*), and in addition such fish as coal fish. Cod is the chief source of raw material, however, and is supposed to make the best grade product.

The fishing area is in the northern section of Norway, with Lofoten as an important center. Fishing is done with hand lines, trawl lines (long lines), and gill nets. The curing stations are usually within a short distance of the fishing grounds. The fish must be especially fresh if they are to make a good quality stockfish. The fish should also be bled as soon as possible after catching. Old or bloody fish will turn sour more easily and are more liable to become wormy, especially if the weather is warm and still.

The first step in the manufacture of stockfish is butchering. This is usually done on land at the curing station but it is done on board if the fishing boats are to remain at sea for more than one day. The process of butchering is as follows: The belly is split open from the pectoral fins to a little below the anal opening, leaving the isthmus in a solid piece. A special knife is used to eviscerate roe cod as the ordinary type damages the egg sacs. The head is then cut on the ventral side as far as the backbone, the cut following a linc just in front of the pectoral girdle. A quick jerk is used to break the head loose from the body.

A second man removes the roe and liver, which are placed in barrels to be saved for later use. He also cleans the fish, removing all visceral material, blood, and other offal. The air bladder is allowed to remain since it is desired by the Italian trade. Stockfish lacking the air bladder is considered of inferior quality in the Italian market. During the summer months all but the very smallest fish are split in two except for a short section near the tail. About two-thirds of the backbone is also removed. Stockfish dressed in this manner is known as "rotskjar" or split fish. "Rundfisk" or round fish are fish that have been beheaded and gutted but have not been split.

After being gutted the fish are thrown into large tanks of sea water for washing. The washing is sometimes not too well done, but Norwegian authorities on fish curing advise that it should be done very carefully, taking care to remove all traces of blood, liver and kidney. If these are allowed to remain a really good cure cannot be obtained as the fish are apt to turn sour and soft.

After washing, the fish are dipped from the tanks to a platform for typing, which is usually done by children. The tails of two fish of about the same size are put through loop of strong twine about three inches in diameter. The fish are turned around two or three times so the loop of cord twists enough to bind the fish. When the fish have been tied in pairs by the tails they are taken to the drying racks in wheelbarrows or special carrying boxes and slung so that the two fish of a pair hang on either side of the drying pole. Two fish hanging in the same loop of twine should not touch more than is necessary. They should not be hung too closely together for good air circulation.

The location and arrangement of the racks is of utmost importance. They are usually rectangular in shape, and of all sizes, from small to large. The structure resembles a gridiron, about 9 feet above the ground. Poles stretch from one side to the other, about 2 or 3 feet apart (fig.3). The racks should be built over dry and airy ground, preferably over bare stone. Consideration of the prevailing winds is important, as the racks should be so set that the wind blows between the rows of fish. A good drying climate is the chief essential in curing stockfish. The air should be dry and there should be a steady breeze. The temperature should not be warm, yet not so cold as to freeze the fish.

At most large drying stations one man is kept busy attending to the racks of drying fish. Each pair of fish is turned at intervals of a few hours, sides that were in being turned to the outside. If two fish are allowed to touch for more than a short period, soft spots will form which may soon become rotten if the fish are not turned. Rotskjar (split fish) are hung individually by the tail in such a way that one-half of the fish hangs on each side of the pole. A two-pronged stick is generally used for hanging up the fish and taking them down.

Thorough cleanliness in and around the drying racks is essential, to prevent damage from blowflies. If any birds, such as crows or seagulls, are around, freshly hung fish should be protected by an old piece of fish netting or some other device. The fish are not considered thoroughly dried if an impression can be made when the thick flesh along the backbone is pressed between thumb and forefinger.

When the stockfish take on a golden color and are so thin and dried along the backbone that the individual vertebrae can be counted, they are ready for removal to storage. Stockfish should be taken from the racks only on a dry day. No moisture should appear on the surface of the fish. If there are any bird droppings, oil, or other dirt on the fish these should be washed off, and the fish allowed to dry afterward. The dried fish are stacked loosely to permit free circulation of air. The doors to the storage rooms are opened in dry weather, otherwise they are closed to keep out all dampness. Most of the fish are shipped in bulk by coastwise freighter to Bergen where they are held in the exporter's warehouse until required for sale.

When the stockfish are to be shipped they are sorted out into piles according to grade. Grades are largely determined by the market, the buyers often sending specifications. In general the product is graded first as to the species of fish or manner of splitting as follows: (1) rundfisk (round fish)—cod; (2) titling—small round cod; (3) rotskjar (splitfish)—cod; (4) hyse—haddock; (5) sei—coalfish; (6) brosme—cusk; and (7) lang—ling. Second, it is graded according to size. Third, the product is graded according to quality. There are two quality grades: (1) Prima or first, and (2) secunda or second. Fish are graded as second class for quality that were old when cured, bloody, not properly split, the air bladder or sound removed, frozen in drying, discolored, wormy, or moldy.

The greater part of the Norwegian stockfish is packed for shipment at the exporters' warehouses in Bergen. The ordinary package is 50 kilograms (110 pounds), but occasionally may be 100 kilograms (220 pounds). Fifty kilograms of the grade desired are weighed out and packed carefully in a hydraulic press. The stockfish are then pressed to about one-third of the volume of the loosely stacked fish. The bale is bound securely with three wires. It is then taken into another room where a cover of sacking is sewn on. The package is stenciled with the grade, the packer's name, and any other necessary shipping marks (Jarvis 1932b).

The principal markets for stockfish are Italy, Spain, the Netherlands, Sweden, and Denmark. The cheaper grades are sold to the Negro trade in Africa. Recently this market has increased in importance and is demanding better grades of fish.

RUNDFISH (COD)	MARKET
Fish per 100 Kilo (220 Lb.)	
100 to 120	Italian
130 to 150	Dutch
180 to 225	Fine Dutch
Titling (small, round cod):	
250 to 350	Dutch titling
400 to 600	Bremer titling

Before cooking, the stockfish should be softened, or disintegrated by beating with a wooden club, and all bones removed. The fish are then soaked for several hours, washed, and drained. In Italy, thin fish are preferred to thick fish and the disintegrated flesh is placed in cold water over a fire, being removed just before the water reaches the boiling point, since boiling toughens the texture. Steaming is a better method of cooking, making the flesh white and soft. In the Scandinavian countries stockfish is made into "lutefisk."

The Russians cure a certain amount of stockfish in the coastal area bordering on northern Scandinavia. The method of preparation is somewhat different from the Norwegian cure. The fish are split through the back and left solid in the belly. A cut about an inch long is made through the uppermost part of the fish and through this the fish is tied up on the drying rack. Stockfish are also prepared in a manner similar to the Norwegian split fish except that the backbone is usually not taken out.

Alaska

Some stockfish is prepared in Alaska. There are two small shore stations in the Shumagin Islands. A few small temporary operations are carried out in other parts of the climate of the greater part of Alaska coast is not particularly suitable. Most of the fish prepared during the winter months when the climate is more favorable. About 12,000 pounds of stockfish were cured in Alaska annually just before World War II. A network of wires is used instead of poles for racks.

Stockfish may be prepared in artificial driers. This has been demonstrated in Halifax with the drier developed at the Atlantic Experimental Station of the Fisheries Research Board of Canada.

STURGEON PRODUCTS

Veziga

The spinal cords of sturgeon are dried for a food delicacy in Russia, Rumania, Iran, and other Asiatic countries where sturgeon are taken in large quantities and, also, to a small extent in the United States on the Columbia River, where this product is being prepared by the Chinese. The spinal cord is removed from the backbone after the head has been removed and the tail cut off. Enough of the spinal cord is pulled out at the tail end with a baling hook to give a good grip for the hands. The spinal cord is pulled out hand over hand until a section about four or five feet long has been removed. This material is cylindrical, whitish in color, and marked or constricted at intervals like a string of link sausages.

It is first washed thoroughly to remove the blood and slime. The cord is hollow and filled with a jelly-like substance which is pressed out by squeezing the cord between the fingers. The larger cords are split lengthwise instead, so that this material may be cut or scraped out and discarded. After cleaning, the spinal cord is washed again in fresh water until it is quite clear in color. It is then hung in a current of air until thoroughly dry.

The whole dried cords are tied up in bundles for marketing or they may be cut in four or five inch lengths which are tied together in small bundles. This product goes by the Russian name “vesiga”, and used to sell for one dollar or more per pound on the Columbia River. The last quoted New York price was five dollars per pound. It takes 25 average-size sturgeon to make one pound of vesiga. The best market is with the Chinese, who use it in fish soups, pies, chowders, and similar dishes.

Balyk

Some sturgeon meat is dried in Russia, where it is known as "balyk". This should not be confused with "beleke", an Alaskan dried and smoked salmon, believed to have been introduced by the Russians. The back flesh only is used in making balyk, the other portions being salted or preserved in some other way. In Russia, the backs of the common sturgeon (*Acipenser guldenstadtii*) and of the "sevriouga" (*Acipenser stellatus*) are cured in a single piece, while those of the large sturgeon (*Huso huso*) are cut either lengthwise only, or else both lengthwise and crosswise. The pieces are placed in a tub so as not to touch each other or the sides of the tub. They are then covered with a thick layer of salt and left from 9 to 12 days. The salt is mixed with saltpeter in the proportion of two pounds of saltpeter to 50 poods (1,800 pounds) of balyk in order to give a reddish color. Allspice, cloves, and bay leaves are frequently added to the brine. The pieces may be left in salt as long as 15 days if they are large or the weather is hot. When the salting is finished, the pieces of sturgeon are soaked in fresh water for about 24 hours to remove excess salt. Then they are dried, first in the sun and then in the shade, on roofed scaffoldings, which are built for this purpose.

The process of drying requires from four to six weeks and is considered finished when the growth of a slight mold covers the balyk. The absence of mold indicates that the balyk has been salted too much. Good balyk must be soft and tender; must have a reddish or orange-brown color; and must also have an odor something like that of the cucumber. It must also be transparent, show no traces of putrefaction, have no bitter taste, and not be too salty. Balyk which meets all of these requirements is considered a strictly fancy grade and sells at a high price. Only a few fish curers are able to make such an article, according to information from Russia.

Miscellaneous Fish

The Chinese who operate the shrimp drying platforms in Louisiana also dry a considerable quantity of fish each summer. Chinese in the San Francisco Bay area also cure each year some miscellaneous fish by air drying. All varieties of salt-water fish are dried, except the small bony ones and the excessively fat. The varieties most commonly used in Louisiana are croakers, red drum, sea trout (squeteague or weakfish), sheepshead, and whiting.

The method used in Louisiana is primitive. Few of the fish are dressed before drying, the heads, scales, fins and viscera remaining, except that the large red fish or channel bass are eviscerated and have the heads and fins removed. In a few instances the fish are given a short brining. The fish are spread out on small slatwork frames, about 8 feet long by 4 feet wide. These frames are set on some type of support and the fish are left to dry in the sun. They are turned every few hours by placing a second frame on the one holding the fish, turning both together and then removing the first frame. Drying requires from one to three weeks, according to the size of the fish and the weather conditions. The product is sold entirely to Chinese markets in the United States.

The fish dried in the San Francisco Bay area are usually made from fish left unsold at the end of the day. They are spread on the roof tops, turned at intervals, and left until dried. Production is now only a small fraction of the former level. With improvement in refrigeration facilities and change in food habits, this product may soon disappear from the market.

Abalones and Other Shellfish

The abalone (*Haliotis* sp.) is prepared on the Pacific coast, where it is a dried product of some importance. At one time, abalones were dried as far north as southeastern Alaska, by the Indians. Now abalones are cured only in southern California and Mexico, by Chinese and other Orientals. The dried product is considered a delicacy by Orientals. Considerable quantities are shipped at times to China, Hawaii, and the Philippines.

Abalones are found at depths from the low water mark up to about 12 fathoms. They are gathered by men in diving suits who pry them loose from the rocks with an implement resembling a tire iron or long-handled chisel. If the iron is slipped quickly under an abalone it is easily detached; but if there is any hesitation the mollusk contracts the muscular foot, and it can only be pried away with difficulty. The diver throws his catch into a net bag or a basket lowered from the boat; he signals when the receptacle is full and the abalones are hoisted aboard. A diver works about three hours at a time.

When the abalones are brought ashore the first step in preparation is to remove them from their shells. The meats are kept for several days in a salt brine testing about 50° salinometer. The purpose of brining is to remove the mantle fringe and preserve the meat during the drying period. When sufficiently brined the abalones are washed, cooked for about 30 minutes in water just below the boiling point, and transferred to shallow trays, where they are spread out to dry in the sun. The meats are turned at intervals during the first part of the drying process. After four or five days the abalones are given another cooking, this time for 60 minutes.

After the second cooking the abalones are dried for a day over a low charcoal fire. They are then rinsed in boiling water and spread out again for the final drying, which requires about six weeks. The dried abalones are rinsed in warm water, wiped, and packed for shipment. The meats lose about 90 percent of the original weight in the curing process. The dried product resembles a piece of hard, tough gum rubber, yet it can be sliced or shaved off in shreds with a sharp knife. The Chinese grind dried abalone to a powder and use it as an ingredient in soups and stews. It can also be soaked and stewed.

A number of other shellfish are dried, mostly in the Orient, with occasional manufacture in the Americas. Considerable quantities of a clam much like our Pacific razor clam are dried in the Orient. These clams are shucked, boiled in salt water about 10 minutes, then spread on bamboo trays to air dry. From two to three weeks are required for drying. The Indians on both Atlantic and Pacific coasts formerly dried quantities of hard clams as a regular part of their winter food supply. Some clams are still dried occasionally by Alaska natives. The clams are shucked, strung on cords, partially dried in the sun, then hung in the rafters of their huts to finish curing over the smoke of the fire. Some oysters are dried occasionally by Chinese shrimp driers in Louisiana. The oysters are handled much like shrimp. Production is almost entirely for use locally though small lots may be shipped to other Chinese settlements in the United States. Mussels are also dried by immigrant fishermen. A New York firm placed a dried clam broth tablet and dehydrated oyster powder on the market about 15 years ago. The principal difficulty with the oyster powder was the high price at which it had to be sold.

DRIED OCTOPUS

Dried octopus is called "pulpo" by the Italians and Greeks. It is also popular with the Oriental trade in this country and a great deal is prepared throughout the Orient. Small quantities have been dried from time to time in the Pacific Northwest and it is cured more or less regularly in California by Orientals for local use. Most of the dried octopus consumed in the United States is imported from the Mediterranean and the Orient.

Levantine Cure

Octopus for the Levantine trade in the United States is prepared by gutting or eviscerating the octopus and washing thoroughly in sea water. The octopus are spread out to dry in the sun on wooden trays or drying frames, elevated a few feet above the ground. As a rule no salt is used unless the weather is unfavorable for drying, or the octopus is not strictly fresh. All sizes are used from the very small specimens dried in a single piece to giants, which may be cut into sections. The drying period may vary from 10 days to 2 weeks, depending on the weather and the size of the octopus.

Oriental Cure

The Oriental cure is much the same as described as above although another variant is sometimes used. After cleaning, the octopus are simmered for about three-quarters of a hour in water just below the boiling point. They are then spread out on drying frames in the sun. A low charcoal fire is sometimes used in drying.

SQUID

Squid are dried extensively in the Orient, especially by the Chinese and Japanese. They are also cured in the Mediterranean where the product is known as "kalamar". While squid are found on both coasts of the North American continent they are dried for market only in California, at present, although they have been dried in Newfoundland. The species utilized is the common squid (*Loligo opalescens*) of the Pacific coast, which has an elongated, pointy body, with triangular fins near the posterior end. This squid ranges from Puget Sound to San Diego. Although a few are caught throughout the year, the period from about April 1 to June 30 is considered the regular squid-fishing season.

The fishing area is almost entirely within the limits of Monterey Bay, with the waters off Moss Landing and Watsonville as important catching areas. Sardine nets such as half-ring nets are probably used most widely, with some lampara nets and a few purse seines. Like sardines, the schools of squid can usually be located only at night in the "dark of the moon" by the phosphorescence of the water resulting from their movements. The only distinction in locating squid rather than sardines is that the former species causes a "fluttery" or "wriggly" phosphorescence in the water, while sardines produce more or less straight lines.

The squid are landed at Monterey within two or three hours after catching, so that no special care in handling is required. They are unloaded at the fresh fish market dock, into wooden fish boxes or piled loose in a deep truck body, then transported to the drying place.

The method used in drying squid is not standardized. It may vary with the curer and according to the quality of the product to be prepared. The best quality squid are first sorted for size. They must be fairly large, from 10 to 18 inches in length. The squid are first washed and split. The quill and sepia bag are removed, and the inside of the body is scraped thoroughly. They are spread out flat to dry, usually on a stretch of hard, bare ground. The squid must be turned at frequent intervals the first few days. As a rule no salt is used in curing. The squid are collected every evening and taken to a warehouse to be protected from night fog and dew. In good weather approximately 10 days are required for drying. When the squid are thoroughly dried and hard, they are packed in barrels to a net weight of about 135 pounds; sometimes in boxes, or matting-covered bales, and shipped to market, usually to San Francisco. It is reported that when properly dried, squid packed in this way will remain in good condition for a year or more.

Second quality squid are not selected so carefully, either as to size or freshness. They are split, washed, cleaned and scraped, salted lightly or brined for several hours, then spread out to dry. They are turned by hand, as are the first quality squid, and require about the same degree of drying. The sale price is somewhat lower.

There is a third grade of dried squid which is distinctly inferior to either of the first two. It is made from the smallest squid, or those taken incidentally in other fishing operations. These squid are not given any preparation, such as cleaning or splitting. No salt is used in curing. The squid are spread on the drying grounds in layers up to 2 inches in thickness. They are turned with wooden or bamboo rakes, as is done in drying shrimp, at approximately 30-minute intervals for the first two days, and at less frequent intervals for the balance of the cure. When dry the squid are shoveled into sacks. The product is then ready for shipment. This quality of dried squid is shipped almost entirely to China where it is sold to the lowest-price market.

Dried squid is called "yu chee" by the Chinese, who consider it a delicacy when properly cured and cooked. Some dried squid is also imported from China. This is a first-quality squid, large, well selected and hard-dried, and sells for about three times the retail price of the best domestic product.

SHRIMP

Sun-dried shrimp probably originated in Asia. The method was introduced into the United States by Chinese, who began a shrimp-drying industry in California some time in the 1860's. Some shrimp are still dried in California, but due to overfishing and the restrictions on the use of the bag net formerly used by California Chinese, production declined to 139,000 pounds in 1941. The drying of shrimp was begun in the Gulf of Mexico area in 1873 by a Chinese named Chin Kee. Production of sun-dried shrimp in the gulf area averages 2,000,000 pounds annually. The development of canning and other methods of preparation has also caused a decline in Louisiana shrimp drying.

Chinese-Americans still handle the marketing of sun-dried Louisiana shrimp, but the shrimp are now cured mostly by "cajun" shrimp fishermen in the Barataria Bay region. Chinese-American fishermen prepare the dried shrimp in California, largely in the San Francisco Bay and Monterey areas. A shrimp-drying camp consists of a warehouse, storehouse, living quarters, some kind of wharf, a platform on which the shrimp are dried, and boilers for cooking. The drying platforms are built near the edge of the water and stand on posts 8 to 10 feet above the level of the ground. They are usually made of cypress. The platforms range in area from 25,000 to 50,000 square feet. One of the largest shrimp drying platforms in Louisiana is 230 feet long by 130 feet wide and has a capacity of about 1,000 baskets of shrimp, or approximately 100,000 pounds. The floor of the platform is not level but has an undulating surface with an ocean-wave effect. These "waves" are about 2 feet in height, with a distance of about 30 feet from crest to crest.

The shrimp are brought directly to the drying place by the fishing boats. The driers utilize the small shrimp, which are not in great demand for canning, freezing, or for sale in the fresh market. The first step is to wash the shrimp and clean them of debris such as seaweed. They are then dumped in large iron kettles holding about 900 pounds. From 10 to 20 one-quart measures of salt are added to each batch, depending on the weather. More salt is required in damp weather than in dry. The shrimp are put in the kettle when the water is actually boiling and the cooking time begins only when the brine again comes to a boil. The cooking time varies from 15 to 45 minutes, depending on the size and amount of shrimp and the weather. The method of determining when the shrimp are sufficiently cooked is to hold one up to the light and observe the shrinkage of meat within the shell. If there is a clear space between the meat and shell the shrimp are cooked.

When sufficiently cooked the shrimp are dipped from the vats into wheelbarrows, drained for about 15 minutes, then taken to the drying platforms where they are spread. The cooking water is changed only about every fifth or sixth lot, but salt is added each time. With long wooden rakes the shrimp are spread out to dry in layers with a maximum thickness of three inches. They are never spread to the maximum depth unless the size of the catch makes it necessary. The shrimp are raked over several times each day to promote drying and prevent spoilage. The frequency with which the shrimp are turned depends on the thickness of the layer, a layer three inches thick requiring attention every 20 minutes.

Each night the shrimp are shoveled and scraped into long low heaps on the crests of the ridges of the drying platform. They are then covered with tarpaulins placed over A-shaped trusses set across the ridges to protect them from rain and dew. The covering must be fixed so as to furnish an opening at each end of the heap to provide for adequate ventilation. If this is not done, the shrimp heat and spoil. When the shrimp are fairly well dried, that tarpaulins may be spread directly on the shrimp, but not in the earlier stages of drying, when ventilation is absolutely necessary.

If the weather is favorable the drying process will require from 24 to 48 hours, depending on the thickness of the layer, the degree of atmospheric humidity and the average size of the shrimp. If rain seems imminent the shrimp are shoveled into heaps on the top of the ridges and are covered by tents formed from the trusses and canvas. The value of the wave-type of construction is now apparent, for even in the hardest rain the shrimp cannot get wet, the water draining away down the hollows. When the sun comes out, the shrimp are again spread out and the drying process continued. If a long period of wet and cloudy weather sets in while the drying platforms are full of shrimp that are still quite green, the whole lot is lost by spoilage. If the drying process is fairly well advanced at the beginning of the rainy period, the shrimp can be gathered and held in the storehouse for a long period. When clear weather returns they can be re-spread on the drying platform to complete the drying.

When the shrimp are thoroughly dried the shells are removed mechanically. Formerly, they were shoveled into large round piles on the platform. The workmen bound sacks on their feet and then trampled the shrimp, going round and round the pile in what may be best described as a slow trot. The process is called "dancing" the shrimp. The trampling separated heads, shell, legs, and other waste parts from the meats. The waste is known as "bran." This method is now used only by a few of the smaller and less progressive packers.

When the shrimp have been well "danced" or shelled, the meats are separated from among the shells by sifting on a coarse wire screen with about 1/4 inch mesh. This does not remove all particles of shell from the shrimp meats. Some firms follow the practice of placing the meats in sacks, which are beaten with boards, after which the shrimp are sifted again. Other driers do not think that the slightly higher price is sufficient to pay for this trouble and pack the meats for shipping without additional cleaning.

Dried shrimp meats are packed in large sugar barrels, to a net weight of about 230 pounds. Some are sold in the United States; in fact, use here is increasing in recent years. The principal market, however, is among Latin and Asiatic peoples. Much dried shrimp is exported to Central and South America, Cuba, China, Puerto Rico, and the Hawaiian Islands. It is very good when used in such southern dishes as jambalayas and curries.

A basket of "green" shrimp, weighing about 105 pounds, should yield about 12 to 14 pounds of dried shrimp meats. The quantity of bran obtained is usually equal to that of the meat.

The "bran" is usually sold locally. It is put up in sacks, about 30 to the ton, and sold either as fertilizer or meal. The shortage of protein materials for stock feeds has led to a study of the use of shrimp bran in animal feeding. Investigators have shown that it has high nutritive value, being an excellent source of both protein and minerals. It is fed to livestock and poultry.

Shrimp drying in California is carried out somewhat differently. It is conducted on a smaller scale, and is probably closer to the original Asiatic method. The shrimp are first boiled in brine. Apparently, the strength of brine varies with the individual curer. When the shrimp are cooked they are dipped into baskets and taken to the drying area. This is a hard-packed area of ground, free from pebbles and vegetation. The shrimp are spread out in a thin layer and turned occasionally with a long- handled bamboo rake. At night they are brought inside. Drying should be completed in four or five days. The shrimp are then piled and trampled on or beaten by wooden clubs to release the meats from the shells. Meats and shells are separated by a crude fanning mill on the same principle as the old-fashioned mills for winnowing grain.

TREPANG

The trepang are holothuroidean echinoderms known in English as sea cucumbers or sea slugs. The dried flesh of the trepang is a very popular food in Oriental countries, where millions of pounds are consumed every year. It is cured extensively in the South Pacific islands, the Philippines and Japan. The sea cucumber is found in large quantities off the coast of Florida, in the Caribbean and the Gulf of Mexico, on the Atlantic coast of the United States. They are found from California to Alaska on the Pacific Coast and are known to be abundant in the Puget Sound area. According to Stevenson (1899) trepang were cured in Florida for a short time. No effort has been made to cure trepang in the United States for many years, though sea cucumbers were canned in the Puget Sound area for several years preceding World War II. The product resembles minced clams.

Seale (1917) reports that there are about 16 principal varieties and 47 commercial grades of trepang in the Philippines, which range from white to black in color. Some are smooth, other are covered with prickles. Live sea cucumbers are from 12.5 to 45 centimeters or more in length, but when dry they are seldom more than 20 centimeters in length and from 2.5 to 8 centimeters in diameter. When properly cured, they look like a bologna sausage and should be dry enough to "rattle like walnuts in a bag." Each species of commercial trepang is divided into three grades; namely, large (toa), medium (tiong), and small (liow), with corresponding values.

The sea cucumbers are taken both by spearing and by diving. To spear sea cucumbers the fisherman usually walks along reefs at low tide. When he sees a sea cucumber it is transfixed on the point of a single-tined spear and transferred to a bag carried at his waist. Sea cucumbers making the best quality of trepang are taken in water from 10 to 30 feet deep and are usually taken by diving. Naked-diving has been the rule but more recently divers have been using helmets. The naked diver drops to the bottom in a favorable spot, fills a basket or bag with the slugs and comes to the top. The diver wearing a helmet works continuously up to three hours, piling up his catch in a container lowered on a rope, jerking on the rope to signal the fishing boat above that the basket is full.

Air Drying

There are several variations in the method of drying trepang, of which the more important are air drying alone, a combination of air and fire drying, and drying entirely over a fire. In curing trepang by the first method, the sea cucumbers are taken ashore as soon as possible and dumped into a pot of sea water. They are split down the side, scraped clean and rinsed and boiled from 10 to 30 minutes, depending on size and variety. The cooked slugs are taken out and spread on racks of reed or bamboo to dry in the sun. The trepang must be turned at frequent intervals during the first part of the cure. Drying takes approximately 20 days in good, clear weather.

Combination Air and Fire Drying

The second method is used generally in the Philippines and is described by Seale as follows:

> "After the trepang are gathered, they are taken to the curing station and are cared for promptly; otherwise they become a blubbery, unsavory mass within a few hours. They are first placed in a pot or cauldron of water (an oil tin would answer the purpose) and boiled for twenty minutes (some require less time). When taken out of the boiling water, they should be hard and elastic and should dry quickly, like a hard-boiled egg. They are slit open with a sharp-pointed knife, and the entrails are removed. They are next placed in the sun and left until almost dry and then transferred to a smokehouse and smoked for about twenty-four hours. The smoked trepang are spread on a mat in the sun until almost dry and then transferred to a smokehouse for about twenty-four hours. The smoked trepang are spread on a mat in the sun until perfectly dry. Finally, they are packed in bags. Trepang are prone to collect moisture, and if kept for three or four weeks they must be again spread out, and dried in the sun to prevent molding."

Fire Drying

In the East Indies, especially on the Malay Peninsula, where storms come up frequently, trepang are dried by fire only. This process is sometimes called smoking, but there is not sufficient smoke to cure the trepang, and preservation depends entirely on the drying action of the fire. The first step is to erect, what is for this area, a large curing house of substantial construction. It has narrow entrances. An average sized house is 90 feet long, 30 feet wide, and 10 feet high. It is covered with a thick thatch or coconut leaf matting, well put on so as to prevent rain from penetrating. Drying platforms are then built along one side of the shed. There are two of these, 8 feet wide, the lowest about 4 feet above the ground,the second about 3 feet above the first. The platforms are generally made with a framework of coconut trees, and a covering of split bamboo or reeds. Under the platforms a trench about 6 feet wide and 2 feet deep is dug, running the entire length of the house. This is used to hold the low fires for drying trepang.

The curing process is as follows: The trepang are first gutted, then boiled from 15 to 20 minutes in large iron kettles filled with sea water. When sufficiently cooked they are washed well with fresh water and carried in small wicker baskets to the curing house, where they are spread out to dry on the lower platform in a layer four or five inches thick. The trench is filled with firewood and lighted, and the drying process begins. From this time the fire must be kept going day and night.

On the first day of curing the slugs must be turned rather often. The next day the fires are put out and the drying trepang are moved to the upper platform. If some are not drying properly, they are transfixed with splints of wood. The lower platform is then filled with freshly boiled slugs, the fires are relighted and the drying goes on.

At the end of the second day the trepang on the upper shelf are inspected, then shifted to one end to make room for another batch from the lower platform. The drying is continued for two more days. The first day's batch should then be cured. It is taken down, examined carefully and packed in bags. Any slugs not completely cured are given additional drying. The fires must always be extinguished when loading or unloading the drying platforms. Unless packed in air-tight containers the trepang must be re-dried in the air at intervals of about 12 weeks as it soon becomes damp again.

The method of fire drying is used most extensively for two reasons: it is much quicker as 4 days are required for curing against 20 days for air drying; and also larger quantities can be handled more readily. However, sun-dried trepang is considered of better quality and brings a higher price.

Trepang preparation is not easy. Gutting must be done carefully, in a large quantity of water, or the trepang putrefies. This also occurs unless curing begins within a very few hours after catching. The trepang blisters and becomes porous if too much heat is used in drying and if there is not enough heat, it putrefies.

SEAWEED

Dried seaweeds are used for human food in many parts of the world, though they are used most often in the Orient. Over a dozen varieties of dried edible seaweeds are sold to the Japanese trade, some bringing three dollars a pound. According to Tressler, "the edible seaweed industries are among the most valuable of the aquatic industries of the Japanese."

The two seaweeds used most widely for edible purposes in the United States and Europe are Irish moss or "carrageen" (*Chondrus crispus*) and dulse (*Rhodymenia palmata*). Irish moss is found along the Atlantic coast of North America from North Carolina to Nova Scotia. This is also the general range of dulse. The Irish moss is gathered commercially at points in Massachusetts, New Hampshire, and New York, and the Province of New Brunswick in Canada.

Irish Moss

Irish moss is gathered from the rocks by means of rakes handled from boats. Some is gathered by hand, especially from the shallow pools near the high water mark. The season for taking sea moss runs from late spring to early fall. If the rocks have not been scraped too closely in the first part of the season, it is said to be possible to get a second crop in some of the warm sheltered coves, where growth is more rapid than in exposed locations.

Good weather with sunshine is essential to curing Irish moss. When brought ashore the weed is washed in sea water, then spread out on the beach to dry and bleach. If the weather is good the Irish moss is gathered after 24 hours and given a second thorough washing in sea water and again spread out to dry. The moss should be completely cleaned and bleached after 3 washings, or about 72 hours, though sometimes it is washed as many as seven times. After the final washing the weed is left in the sun until it is completely dried. This usually requires about two weeks of good weather with sunshine. The Irish moss gradually loses its color in this drying period, finally becoming almost white or light straw color.

Drying is an operation requiring great care, and rainy weather can easily spoil the Irish moss. The weed must be raked up and covered at the first sign of rain. Irish moss is used chiefly for making blanc-mange, other puddings, and jellies. It is also used as a stabilizing agent in foods and beverages, as for example salad dressing and chocolate milk.

Dulse

Dulse is particularly abundant from New England to Nova Scotia. It is found on rocks near the shore. The process of preservation is very simple. After washing in sea water to remove sand or other debris, the dulse is spread on rocks or shed roofs to dry in the sun. Much of the dulse consumed in New England is imported from Nova Scotia, in light wooden boxes holding about five pounds, or in sugar barrels shipped loose. For retail sale the dried weed is packed in small cartons holding about four ounces, like those used for shredded codfish. It is eaten as a relish and is used in stews as a thickening and flavoring agent. The Scotch and Irish sometimes chew the dried weed instead of tobacco. It is valued for its high iodine content and is used in the prevention of goiter.

Chapter Five

Dehydration of Fish

DEFINITION AND PRINCIPLES OF DEHYDRATION

There is some confusion in applying the word "dehydration" to fish-curing methods. The term is sometimes used loosely. The following definitions have been suggested by D. L. Cooper after a discussion on the subject with the writer.

> "Dehydration is the drying of fish protein, with or without the addition of salt, without respect to form. Drying is the dehydration of fish, with or without the addition of salt, in which the structure and form are retained."

According to Tressler (1923) dehydration preserves fish by destroying enzymes and removing the moisture necessary for bacterial and mold growth. Fat fish cannot be dehydrated by the ordinary dehydration process, nor is is it possible to store dehydrated fat fish in the usual way. Fish oils or fats are drying oils which rapidly absorb oxygen from the air and harden just as paints harden on exposure to air. Fatty fish must be dehydrated quickly in a vacuum, and must be stored *in vacuo* or in an atmosphere of an inert gas.

PREVIOUS WORK

Experiments in the dehydration of fish have been conducted over a period of 70 years. Most of the early experimental work was haphazard, and none of the methods developed was successful. Stevenson (1889) describes some of the early efforts, which were tried at the same time on fruits and vegetables.

The dehydration of fish has been studied in Germany and Norway for many years, and it is reported that scientific methods were applied successfully as early as 1913. There are a number of references made to the subject in German technical journals. Tressler (1923) describes a process for the dehydration of fish put into commercial operation in Cuxhaven, Germany, about this time. Though references from German sources claim the development of successful methods for the dehydration of fish, so far as can be determined there was little or no dehydrated fish used during the war period of 1939-45.

STUDIES IN WORLD WAR I

At the time of the First World War a comprehensive study was made in the United States by Scott and Wolff (1936), under the direction of the United States Bureau of Fisheries, and the data were summarized by Tressler (1923). This investigation is the first scientific study reported in English on the dehydration of fish. These workers found that raw fish did not dehydrate well; that fish must be pre-cooked to obtain a dehydrated product of satisfactory quality. They determined also that the cooked fish must be flaked and ground before dehydration. It was also concluded that moisture content should be reduced to one percent of less to avoid changes in protein after dehydration. Dehydration *in vacuo* gave the best results, but this method was considered too expensive. Oxidation in storage was a serious problem. Light was also found to cause deterioration in the stored product. The best results were obtained with non-fatty fish packed in hermetically sealed containers and stored in a cool place.

The method of Scott and Wolff in *Marine Products of Commerce* required that the fish be cleaned in the customary manner. Heads, viscera and fins are removed and the fish are scaled. They may be skinned either before or after pre-cooking; as a rule it is done more easily and quickly after cooking. The cleaned and washed fish are placed in shallow pans with perforated bottoms and cooked in a steam chest under pressure. The process used was 20 minutes at 12 pounds pressure (244° F.). The flesh is picked from the bones of the cooked fish as soon as possible. The flaked fish are run through a grinder, using a plate having perforations three-sixteenths of an inch in diameter.

The ground fish are spread on shallow drying trays in a thin layer. The loaded trays are placed in driers and the moisture is removed by currents of warm air. The temperature in the drier is about 145° F. at the start. As the process continues, the temperature is gradually reduced, because too high temperature gives a noticeably discolored product. The temperature of the fish must not reach that of the air in the drier. The drying is discontinued when the moisture content is reduced to five percent. This requires about two hours in a good warm-air drier. The dried product is ground coarsely and packed in air-tight containers.

Interest in the dehydration of fish was not great in the period following World War I. Some work was done in Norway on the production of a "fish flour" but the product was not accepted. A study on fish dehydration was conducted at the College of Fisheries, University of Washington, to test Scott's method and to determine what North Pacific coast species might be utilized. It was found that lingcod, (*Ophiodon elongatus*), Pacific cod (*Gadus macrocephalus*), rockfish (*Sebastodes* sp.), Pacific halibut (*Hippoglossus stenolepis*, octopus (*Octopus punctatus*) and squid (*Loligo opalescens*) were suitable for dehydration.

A private firm in the United States produced two dehydrated shellfish products, an oyster tablet and a clam broth tablet which were put on the market during the 1930's. These products were of excellent quality but did not obtain a very wide distribution commercially.

STUDIES IN WORLD WAR II

The outbreak of the war created many problems in the transportation, preservation, and storage of foods. As a result there was renewed interest in dehydration. Much progress was made in the dehydration of fruits and vegetables. Products of high quality were developed and placed in large scale production. Some of these are having a reasonably good reception as commercial products. There was also interest in the possibilities of other dehydrated products, including meat and fish.

U.S. FISH AND WILDLIFE SERVICE

In 1942, when shipping space was extremely scarce and it was also feared that tin supplies might soon be exhausted, military and other government purchasing agencies indicated much interest in the possibilities of a dehydrated fish of good quality. As the problem fell within its sphere, the U.S. Fish and Wildlife Service included studies on dehydration in its war problems research program at its technological laboratories in College Park, Md., and Seattle, Wash. The work in the Seattle laboratory was principally engineering studies on the development of a satisfactory product and method of manufacture, with investigations on the effect of various processing methods, storage conditions and types of containers on the storage life of the product. At the College Park laboratory, attention was given largely to the suitability of different species for dehydration, the determination of the nutritive value of these species when dehydrated, and the possibilities of several new and untried types of apparatus in the dehydration of fish.

A horizontal tunnel drier in which temperature and relative humidity were automatically controlled and air velocity could be varied between zero and 900 feet per minute was used in the engineering studies. The drying section was 18 feet long, with cross sectional dimensions of 18 by 23 inches. It was provided with four racks, each holding seven trays with an area of 2 1/2 square feet.

The investigators found that fish could be dehydrated successfully over a rather wide range of temperature conditions. Stansby (1944) reports that:

> "...drying temperatures as high as 205° F. can be used in the initial drying stage without harmful effect. As the product becomes drier, it is necessary to use a lower temperature in order to prevent scorching and in the later stages temperatures above 145° F. are inadvisable. Relative humidities between 10 and 40 percent have no noticeable effect on the quality of the product. Low humidities and high initial drying temperatures are, of course, helpful in increasing the rate of drying."

This is in contrast to the earlier results of Scott (Tressler 1923) and the results of recent studies by the British Food Investigation Board (Cutting 1944) which give a maximum temperature of 70° C. (158° F.).

The staff of the College Park laboratory of the Fish and Wildlife Service cooperated with a number of manufacturers of industrial dehydration equipment used for other purposes to determine the adaptability of certain new apparatus. Both the Megatherm process utilizing radio-frequency energy as a source of heat and one employing high vacuum desiccation of the frozen material were given a preliminary trial. Results obtained from both types were promising, but since at the time neither was sufficiently developed for commercial use in the dehydration of fish, no attempts were made to work out a commercial process. Compression of dehydrated fish into small blocks was also given some attention at the College Park laboratory. Dehydrated fish were stored at various temperatures from 40° to 100° F. to determine the effect of varying temperatures (Young and Lee 1943).

SPECIES SUITABLE

Species considered suitable for dehydration include Atlantic cod (*Gadus morhua*); angler fish (*Lophius piscatorius*); Atlantic croaker (*Micropogon undulatus*); groupers (*Epinephelus* sp.); striped mullet (*Mugil cephalus*); puffer (*Spheroides maculatus*); sea robins (*Priontus* sp.); gray sea trout (*Cynoscion regalis*); rajafish or ray (*Raja* sp.); whiting *(Merluccius bilinearis*); and sea mussel (*Mytilus edulis*) from the Atlantic coast; burbot (*Lota maculosa*) and carp (*Cyprinus carpio*) from inland waters; and Pacific cod (*Gadus macrocephalus*); rockfish (*Sebastodes* sp.); Pacific halibut (*Hippoglossus stenolepis*); Pacific herring (*Clupea pallasii*); pilchard (*Sardinops caerulea*); petrale sole (*Pleuronectes* sp.); chum salmon (*Oncorhynchus keta*); king salmon (*O. tschawytscha*); silver salmon (*O. kisutch*); Pacific mackerel (*Pneumatophorus diego*); and squid (*Loligo opalescens*) from the Pacific coast.

STORAGE CHANGES

Storage changes taking place in dehydrated lean fish include development of a tough texture, darkening of color, and a burnt flavor and odor. Stansby (1944) states that:

> "These changes are identical with those encountered with dehydrated eggs where such changes are attributed to a chemical combination between the protein and certain sugars which occur only in small traces in eggs. With

> dehydrated eggs these changes are minimized by reducing moisture content to less than one percent. It seems possible that a similar improvement might be obtained by dehydrating fish to a similarly low level."

He concludes that unless the keeping quality of dehydrated fish can be greatly improved there is apparently no possibility of a domestic market for the product.

Cost of Production

Both Stansby and Young have made some estimates of cost of production. Stansby arrives at a figure of $1.50 a pound of finished product for cost of the raw material exclusive of all processing and overhead charges where haddock is used when the price paid to fishermen was about 30 cents a pound. Young made a more detailed estimate, including all factors in the cost of production such as labor, fuel, and overhead. His raw material costs are lower but the total cost of production is estimated at $1.80 a pound so that the price of the wholesaler would be approximately $2.00 a pound. Both conclude that even if "trash" fish were used, the cost of manufacture would be so high that the average consumer would be unwilling to purchase dehydrated fish. It is concluded that at present the outlook for development of a commercial dehydrated fish industry on a permanent basis, either for domestic or foreign markets, is not promising.

BRITISH MINISTRY OF FOOD

The most extensive studies of recent times on the dehydration of fish are those of the British Ministry of Food, Department of Scientific and Industrial Research, at the Torry Research Station, Aberdeen, Scotland (Cutting and Reay 1944 and Cutting 1944). This agency is also the only one which has placed dehydrated fish in actual consumption on any scale, so far as is known at present.

In the course of the experimental work at the Torry Research Station eight types of "lean" fish were used, including five species belonging to the family Gadidae, and three types of "fat" fish; Atlantic herring (*Clupea harengus*); mullet (*Mugil* sp.) and Atlantic Salmon (*Salmo salar*). These workers also came to the same conclusions as Scott, namely that experiments in drying raw fish, either in fillets or ground, showed this procedure to be unsatisfactory. This was also true of whole cooked fillets. Cooked whole fish were not acceptable. Gadoid fish and herring were dried both fresh and cooked, and smoke-cured before cooking. The smoke-cured herring are believed to make a more acceptable product than the fresh cooked. While dried "white fish" made a product with longer storage life, large-scale operations have been devoted mostly to the dehydration of herring for reasons of cost-price and availability of raw material.

Description of Drier

The drier used (British Patent 544,229) is of the tunnel type developed from the controlled smokehouse invented at the Torry Research Station before World War II.

> "It consists essentially of a chamber holding the four tracks. A 35-inch axial flow fan driven by a 10 hp. motor, is fitted into a recirculation duct over the chamber. Two steam heating batteries are provided, one in the recirculation duct and a smaller one in the middle of the chamber between the second and third tracks. The drying chamber is fitted with adjustable diffuser walls at inlet and outlet ends which ensure uniform air-flow over any cross-section. This provision was the special feature of the original smoke-curing kiln. The fan delivers about 30,000 c.f.m. Fresh air is taken in through adjustable louvres placed before the heaters and behind the fan in the duct, and wet air is vented through an exhaust duct at the bottom of the chamber just below the outlet diffuser wall, an adjustable flap giving venting control. The steam heating is controlled by thermostats operating motorized valves. The drier is insulated with magnesia block and is satisfactorily air tight (Anon., undated.)."

Description of Method

The manufacture of fresh-cooked dehydrated herring as developed for large scale operations at the Torry Research Station is as follows: As soon as the fresh herring are landed they are packed in quarter-cran boxes (about 98 lb.) and well iced. The boxes are delivered to the plant by truck. The first operation at the plant is washing and scaling. The machine used is essentially an inclined drum, made of rods spaced closely together, and revolves in a tank half filled with water. As the fish are carried through the drum, the tumbling action and the abrasive effect of the rods forming the drum wall remove all slime, scales, and other debris.

The washed fish are mechanically split, gutted, "block" filleted and given a second washing, using special machines each with a capacity of 1 1/2 to 2 crans of fish per hour. The report of the Torry Research Station states that: Four machines are installed, which should easily handle expected full daily capacity of the plant, amounting to about 28 crans required for two drying runs per day. Roes and milts are canned; the guts, bones, and other waste are sent to a fish meal factory. The washed fillets are spread on sheet metal trays with perforated bottoms, in an amount of about 28 pounds per tray. The individual trays are 29 by 33 inches. The loaded trays are placed in wheeled angle-iron racks which hold 11 cooking trays spaced vertically at 2 1/4 inches apart. The racks are run into a horizontal retort of the Portland type. A retort will hold two racks. The fish are cooked for 20 minutes at 6 pounds steam pressure (230° F.). The racks of cooked fish are wheeled from the retort to a small tunnel cooler where they are cooled in a current of air at surrounding temperature until the fish reach a temperature below 100° F., which requires about 30 minutes.

When cooled, the cooked fish are emptied into a hopper from which they are fed to a power-driven grinder, using either a 1/4 or 3/8 inch plate. The ground material is spread in thin layers on drying trays with tubular metal frames and bottoms of heavy galvanized wire mesh. Each tray is loaded to an average of 2 1/4 pounds of ground material per square foot of tray area, each tray load being check weighed. Loaded trays are placed in wheeled dryer racks. Each rack has a capacity of 44 drying trays spread 2 inches apart. The dryer holds four racks, the total load of cooked fish being about one ton.

The operation of the dryer is described as follows: The drier is warmed up and after loading with four trucks it is set to run with a supply air temperature of 175° F. with intake and outlet vent open full. The wet bulb temperature rises quickly to about 125° F., then falls gradually thereafter. After 1 1/2 hours the dry bulb temperature is set at 160° F. and held there for the rest of the run which takes about 4 1/2 hours. At the end of this time the dried material, about 900 pounds, having a moisture content normally reduced to 2 to 4 percent is cooled and stored in large covered metal bins for packing.

The dried fish is packed in No. 10 cans, each holding 3 pounds. To obtain this fill it is necessary to press the fish lightly with a hand-operated pressing machine. The filled cans are sealed by a hand-operated closing machine. The cans are then nitrogen filled, using the same method as in the British Ministry of Food vegetable dehydration plant. The cans are brogued, that is, punctured with a small hole, then placed in gas packing cabinets. Each cabinet has a capacity of 12 No. 10 cans, with wooden partitions. A vacuum pump is connected with the closed cabinet and the air is exhausted until a minimum pressure is reached. This usually requires about 2 minutes. Nitrogen is then admitted steadily, the pressure being raised to 2 pounds per square inch above atmospheric pressure. The "brogue" holes are soldered immediately after the cans are removed from the cabinet. When soldered, the cans are given a coat of outside lacquer and are packed in wooden boxes, six to the box.

Chapter Six

Principles of Fish Salting

Salting as been the most important method of fish preservation since prehistoric times. In the United States salting has been displaced almost entirely by canning and freezing. In Canada, these methods now equal salting in production, but in all other countries with fishery industries the quantity of fish salted is far greater than the production of canned or frozen fish.

Salt is used in almost all methods of fish preparation, either as a condiment, flavoring agent, or as an accessory preservative. Many dried or smoked fish are lightly salted or at least flavored with salt, but speaking accurately, only when salt is the principal preservative should the method be called salting.

PRESERVATIVE ACTION OF SALT

Salt is not an antiseptic in the accepted sense of the word, though it is a valuable agent in the prevention of spoilage. Salt preserves by extracting water. If fish are laid in salt, the salt soon removes sufficient liquid from the fish to form enough brine to cover them. At the same time the water is being removed from the flesh, salt enters the tissues of the fish and soon the body juices have become a concentrated salt solution. When enough salt has entered to coagulate all the proteins that are coagulable by sodium chloride and when the tissue cells have shrunk because of the loss of a large share of the moisture content, the fish flesh loses most of its translucent appearance and does not feel sticky to the touch. At this stage the salter would say it is "struck through."

The action of the passage of the salt into and the moisture out of fish flesh is an example of the phenomenon of osmosis. In the osmotic action of salt solution with the juices of the fish flesh, the skin and cell membranes act as imperfect semi-permeable membranes. Dilute cell liquid, mostly water, passes from the fish and the entry of some salt into the cells precludes the withdrawal of the colloidal protein of the cells. The direction of flow of liquid through a semi-permeable membrane is always from the weak to the stronger solution. Therefore, when fish are placed in strong brine, liquid passes rapidly out of the cells of the fish flesh, through the cell wall into the brine. On the other hand, salt from the brine slowly passes through the cell walls into the protoplasm. Ultimately a state of equilibrium is reached in which the concentration of the solution inside the cells of the fish tissue is equal to that of the curing brine. When this point has been reached, the salting action is complete.

FACTORS AFFECTING THE SALTING OF FISH

There are a number of factors affecting the salting of fish which determine the quality of the finished product. The length of preservation is dependent to some extent on variations in these factors. The most important are outlined briefly here. The subject is treated at length by writers cited in this report.

Composition of Salt

Tressler (1923), Boury (1932), Carter (1932), and Reay (1936), have found that the composition of salt is of great importance not only in affecting rate of penetration, but also in determining the physical nature of the finished product. Common salt is a simple chemical compound composed of sodium and chlorine; its chemical name is sodium chloride. It occurs widely in nature; in the sea and in deposits in the earth. Salt is obtained by mining or the evaporation of sea water. Salt used for food purposes, however, is never chemically pure sodium chloride as it contains small amounts of other minerals and organic material as impurities, and a small amount of moisture. This is especially true of "solar" salt, that is, salt obtained from the sea. The principal impurities in ordinary commercial salt are calcium salts, such as calcium sulphate; magnesium salts, for example, magnesium sulphate, and organic matter. Solar or sea salts are used most widely for the salting of fish, though more mined salt is being used since Tressler reported his study on the subject in 1920 and 1923.

Salts of calcium, magnesium and the other sulphates retard the rate of penetration of common salt into the fish flesh during the curing process. This is more likely to permit decomposition of the protein of fish flesh during the salting process. Therefore, it is important to use the purest salt obtainable when the weather is warm, or other adverse salting conditions. The U.S. Bureau of Fisheries (now the Fish and Wildlife Service) in a series of experiments measured the rate of penetration, using different types of salts (Tressler 1920). In this series were included chemically pure salts and salts containing various impurities in different known amounts. It was found that a salt containing 4.7 percent magnesium chloride, penetrated no farther in 5 days than pure salt did in 3 days. This is definite proof that a salt free from impurities should be used if rapid penetration of fish flesh is desired.

There is yet another difference in the effect of pure and impure salt in fish salting. It has been found that calcium and magnesium salts have a marked effect on both the color and texture of the salted fish. Fish salted with pure salt are limp, soft and have a straw or cream color. Such fish are more easily freshened than the usual commercial salt fish and resemble fresh fish more closely when cooked. If ordinary fish curing salt is used, the texture is much more stiff and brittle and the color is the chalky white usually associated with commercial salt fish of the cod group. It has been found that the presence of as little as one percent of calcium or magnesium in salt causes a remarkable whitening and stiffening of the flesh. Since impure salt has always been used in salting, buyers demand a firm white-fleshed fish. It is believed that the fish cured with pure salt is in every way superior, but the curers must be educated to prepare fish of this description, and the public to buy it. Salts of both calcium and magnesium give the strong, acrid flavor characteristic of commercially salted fish.

Influence of Temperature

Fish are salted over the temperature range of 32° to 100° F. It is the common experience of fish-salters that at the higher temperatures there is very rapid "striking through" or penetration of salt, much faster than at normal temperatures of about 60° F. Fish may readily become oversalted at higher temperatures unless they are dried after a shorter brining time than usual.

It is asserted by Taylor (1922) that cold, approximating the freezing point, also promotes rapid penetration of salt and it is claimed that this has been definitely proved by experiment. On the other hand, unpublished work on the salting of salmon conducted at the College of Fisheries, University of Washington, indicates that the rate of penetration of a 90° salinometer brine at 32° F. is two-thirds the rate at 60° F. A practical proof of the University of Washington data is found in the commercial mild curing of salmon. If the sides of salmon are placed unchilled in the curing brine, they absorb too much brine, which also penetrates too rapidly so that the flesh is over salty and improperly cured. Therefore, the sides are chilled for 2 to 4 hours in a tank of iced 30° to 40° salinometer brine before salting so that the rate of penetration will be delayed. It should also be noted that 39° F. is considered the best temperature for mild curing. Scofield (1925) states that a temperature above 44° F. is not safe for mild curing salmon.

Influence of Method of Cleaning

The preliminary treatment, or butchering and cleaning, varies with the species and size of the fish and the type of product demanded by the market. Some fish, such as certain types of herring, are not cleaned at all, but are salted round. In salting Dutch cure and Scotch cure herring the fish are cut at the throat and the gills removed so that the blood flows freely and is collected in the brine. The bloody brine gives a distinctive flavor much desired in certain European and American markets. Other small fish are beheaded, split down the back and thoroughly cleaned. Large fish are split into sides or fillets, with all blood, viscera, and other offal completely removed. Salted fish prepared by the last two methods do not spoil so quickly as the first two.

The studies of Tressler (1920) on the influence of different methods of cleaning as applied to the rate of decomposition in salt fish, have indicated clearly that it is essential to remove all blood and offal if decomposition is to be reduced to a minimum. Tressler made a series of experimental packs of salt fish in which fish were cleaned by various methods, and in greater or less degree. Two complete series were made, one held at 79°, the other at 88° F. during the salting process. The amount of amino nitrogen developed was determined by analysis. It was found that the amount of amino nitrogen increased in proportion as the degree of cleaning decreased. The development was greatest at the higher temperature. Only the fish which had been thoroughly cleaned and freed from all blood were salted successfully at temperatures of 79° and 88° F. While high temperatures were chosen for the test to provide a more striking illustration in the shortest time, results will be the same at a lower temperature though the time required will be longer.

Decomposition may occur in blood or flesh, or in both. Since blood decomposes rapidly it is believed that in the work just described, only the blood decomposed and pervaded otherwise sound fish giving a tainted odor and poor appearance. If fish are to be salted in warm weather it is necessary that the blood be removed thoroughly. This cannot be done by eviscerating the fish with the average amount of care and rinsing in water. The diffused blood must be leached from the flesh and veins. The false kidney which is the dark bloody streak just below the backbone must be scraped out and the fish thoroughly washed in clean water.

Influence of Method of Salting

There has been some disagreement as to the relative merits of dry and brine salting of fish. Northern curers generally use the first method while salters in the Middle and South Atlantic States favor the second. For the purpose of this discussion, dry salted fish are those covered with dry salt and the brine is allowed to form naturally. In brine salting the fish are emptied into tanks containing enough brine to cover them, instead of waiting for the brine to form naturally. Southern packers claim that it takes too long for the brine to form naturally and that deterioration occurs unless the fish are covered immediately with brine. Tressler (1920) also determined the comparative efficiency of these two methods. From his work and that of other investigators it has been determined that brine-salted fish consistently undergo greater decomposition than those cured in dry salt. Tressler found that the average excess of amino nitrogen in six lots of fish pickled in brine over that of six lots cured in dry salt was 51 percent.

Fat fish cannot be exposed to the air in curing. Such fish as cod can be dried in the air either with or without salting because the fat content of the flesh is low. Salted mackerel, herring, or salmon must be kept away from contact with air. If they are not so protected, "rusting" sets in; that is, a brownish discoloration appears on the exposed surfaces and rancidity develops. The fish are soon no longer marketable. Every year much salted mackerel and herring are lost through rusting for this reason.

Because of the development of rancidity and as fat in any appreciable amount retards the rate of salt penetration, it is difficult to salt fat fish successfully and store them over a fairly long period of time. To do so, they must be kept away from both light and air. The use of chemical agents is possible in certain fields to prevent oxidation and rancidity. The Fish and Wildlife Service has done some work on the use of chemical preservatives or reducing agents in preventing rusting and oxidation. None of the agents tested was found to be very effective in the concentrations ordinarily used in other foods. About all that can be done is to make sure that the fish are in tight barrels, are completely covered with strong brine, and held under proper storage conditions.

In addition, the comparative rate of penetration of the salt was determined. It was found that the percentage of salt was higher in both the inner and outer layers of the flesh in dry-salted fish than in brine-salted fish. The rate of penetration of salt is about 20% more rapid in dry-salted fish. This is believed to be due to the fact that in dry weather there is always an excess of salt so that the brine is not permitted to fall below the saturation point. This is to be expected in accordance with the effect of osmosis. The flow of liquid in osmosis is always from the less to the more concentrated solution. Brine is weakened rather rapidly by the extraction of water from fish tending to bring the solutions inside and outside the fish into equilibrium and therefore slowing the curing process.

Behavior of Fat in Salting

Fat fish like herring cannot be exposed to air in curing. If so, "rusting" sets in and rancidity develops. Because of this and because fat retards the rate of salt penetration, it is difficult to salt and store fat fish successfully. All that can be done is to pack the fish in tight barrels, completely cover them with brine, and hold in chill storage. The Fish and Wildlife Service has studied the use of chemical preservatives in salting fat fish. None of those tested was found to be very effective.

Influence of Storage Conditions

The kind of storage is a deciding factor in determining the length of preservation of salt fish. The importance of storage in canning has been stressed elsewhere by the writer (1943) but proper storage is even more essential in fish curing. Salt fish is not protected by a hermetically sealed container. It is liable to more types of spoilage than canned fish. Improper storage may greatly reduce the normal length of preservation of salt fish.

The familiar injunction to "store in a cool, dark place" has been used almost as long as foods have been preserved commercially, and is generally conceded to be a necessary admonition. The reasons for this instruction are among the most important points in successful food storage. In most chemical reactions the speed of the reaction increases rapidly with a rise in temperature. It has been calculated that the speed of reaction is doubled with each increase of 18° F. In experiments conducted at the Technological Laboratory of the Fish and Wildlife Service at College Park (Md.) samples of herring stored at 95° F. spoiled twice as rapidly as samples held at 80° F. Salted river herring stored at approximately 40° F. have been held in good condition more than a year whereas herring held at room temperature, about 70° F., spoiled in 90 days. Light, or more particularly sunlight, by itself also hastens the oxidation and hydrolysis of the fat of fish. There may also be a physical softening of the tissue structure. Therefore, salt fish should be stored in a cool, dark place especially if it is fat.

Herring, mackerel, salmon, and any other fat fish are usually held under the surface of strong brine, in chill storage. On the other hand, the enzymic or autolytic deterioration of salt fish flesh is so slow under proper storage conditions that fish will remain in good condition for at least two years. The importance of proper storage in preventing reddening is discussed at length in the section on reddening under the title, Spoilage in Dry Salt Cod and Other Fish.

Chapter Seven

Dry Salting Cod

On a world basis dry salt cod is second only to herring in importance as a cured product. In this discussion the term "dry salt cod" will include the flesh from various members of the family Gadidae cured by dry-salting. The fish of this family are taken in the same areas, cured by the same method, and the final product is similar. It is also customary in the salt fish trade to include these species with dry-salted cod. Species of Gadidae other than cod are of minor importance as a salt product. These species are also classed together as "groundfish" since they live on or near the bottom. When salted or dried, the groundfish other than Gadidae are generally known as "scale fish."

SPECIES AND RATE OF PRODUCTION

The various species of Gadidae salted in the United States and the estimated proportion they form of the total amount salted are: cod, Atlantic (*Gadus morhua*) and Pacific (*Gadus macrocephalus*), 94 percent; haddock (*Melanogrammus aeglefinus*), 1 percent; pollock (*Pollachius virens*), 3 percent; hake (*Merluccius bilinearis*), 1 percent; and cusk (*Brosme brosme*), 1 percent.

The salting of cod is widely distributed through the north temperate zone. Norway, Newfoundland, Iceland, Canada, the United States, the United Kingdom, France, Portugal, and Spain are the principal sources of salt codfish in about the order named. Unknown but probably important quantities of salt fish are produced in the Union of Soviet Socialist Republics. Some dry salting of cod is also done in Germany, the Netherlands, Sweden, Denmark, and Japan.

The salt-cod industry was at one time the most important fish preservation industry in the United States. At one time more than 80 percent of the catch was salted. The poundage salted and its importance in relation to the other fishery industries have both decreased greatly in this country. In 1914 the amount of salt cod produced in the United States was 68, 456, 917 pounds. In 1919 the amount was 62,902,037 pounds; in 1929, 28,973,492 pounds and in 1940, 17,737,263 pounds. Domestic consumption has not decreased to the degree indicated by the production figures, since the United States has become an importing instead of an exporting country. Imports of salted cod (groundfish, all types) supply about 80 percent of domestic consumption. Imports in 1939 totaled 53,918,000 pounds. Approximately one-half the imports of salted cod or about 23 million pounds are consumed in Puerto Rico and the Virgin Isles. About 95 percent of the imports originate in Newfoundland and Canada.

The decline in the salt-cod industry of the United States is due to a number of factors. The great increase in the production of canned fish has absorbed a portion of the market. The development of refrigeration and the filleting industry has not only diverted a large part of the raw material formerly used for salt cod, but has also taken part of the market. Another factor is the great increase in the price of raw material and other production costs, so that it is no longer an economic method of preservation in the United States, except under special circumstances. Change in food habits is another factor.

GROUNDS WHERE CAUGHT

Cod and related groundfish are found on banks which are submarine plateaus on the Continental Shelf where the water is cool and rich in food material. At certain seasons cod can be found in bays and inlets close to shore. Then they go off to the banks again. There are two types of banks, "inshore banks", usually small, close to shore or between the outer islands; and "offshore banks" found farther out to sea and generally large in area. The Grand Banks of Newfoundland are a good example of the offshore bank. Cod are found at depths of up to 100 or 150 fathoms. Beyond these limits the bottom drops off rapidly to one or two miles, when conditions are otherwise unfavorable. The banks on which cod are found stretch from Long Island to the coasts of Greenland on the Atlantic coast of North America. They amount to at least 260,000 square miles in area which is equivalent to the area of all the coastal States from Maine to North Carolina.

Cod are found on the Pacific coast from Puget Sound along the coast of British Columbia and Alaska extending into the Bering Sea almost to the island of St. Lawrence. The Pacific cod banks are reported to be larger and of greater potential production than the Grand Banks of Newfoundland. There are, too, extensive banks off the Siberian coast.

Gear And Methods Of Catch

Various types of gear are used in taking cod. The otter trawl is most important in the United States fishery, but nearly all of this catch is sold fresh. The trawl line and hand line are the traditional types of apparatus used by salters in the United States, Canada, and Newfoundland. Floating traps, pound nets, and sunken gill nets are also used. Objections made to the otter trawl as gear in catching cod for salting are that the fish are taken in such large quantities that they cannot be handled speedily enough to prevent deterioration before they can be salted. This objection is also made to the use of the trap. The best quality fish are reported to be those taken by hand line.

Handling Raw Material

Handling of the raw material before it is brought to the plant and placed in cure, largely determines the quality of the finished product. For the best quality product, the fish should be alive when taken from the gear and bled immediately by the fisherman. This may be done by cutting the throat, pulling out the gills, or cutting out the tongues. Some fishermen cut out the tongues as a means of keeping count of the catch and for personal consumption. Bleeding is recommended because diffused blood in the flesh darkens and discolors the tissue, and deteriorates much more rapidly than the flesh itself. It is, therefore, an excellent medium for bacterial growth. Bleeding is, of course, impracticable in handling otter trawl or trap catches but it should be used wherever possible. The Norwegians are most careful about bleeding fish, and this practice is followed by some Canadians, but rarely if ever by curers in the United States.

Some cod has already deteriorated in quality when it is taken from the water because of the type of gear used and the method or lack of method of operation. Weather conditions sometimes affect the operation of gear, and these, of course, cannot be controlled. Since gill nets strangle fish in the act of catching them, the fish have already been dead for some hours when taken from the water, reducing by so much, the time that may safely elapse before they are put in salt. Fish may be left on trawl lines for two or three days especially when stormy weather prevents overhauling the lines. Cured fish from this raw material will have a darker color and the flesh is apt to be broken and loose, especially near the backbone.

Rough handling on the fishing vessel increases the rate of deterioration, making it more difficult to prepare a well-cured product. This is because bruising and crushing of the fish greatly increases the rate of autolytic decomposition. Fish should be carried or transferred by conveyer, not thrown about; it is a wasteful practice to use the pew or one-tined pitchfork to toss fish from the boat to the wharf or into the unloading bucket. The use of the pew or fork is favored by some fishermen because it is an easy method of handling, but it also encourages carelessness and pitching, with consequent bruising and crushing, thereby lowering the quality of the raw material. Also important is the fact that the flesh is pierced by a number of holes. These holes are excellent inoculation centers for bacteria, so that tissue around the holes is soon darkened, and may acquire an off flavor before other portions of the fish. The use of pews should be prohibited as much as possible. If they cannot be dispensed with entirely, the curer should insist that fish be forked only through the head, and laid, not tossed, on the wharf. However, it is advisable to land the fish by other means, such as by hoist or elevator conveyer.

If the fish cannot be put in salt within three or four hours they should be well cleaned, the gills pulled out, and they should be iced. The amount of ice depends on the season; a good general rule is that the fish should be well covered. The cleaning usually given on the fishing boat is apt to do more harm than good. The blood is not scraped out and bits of viscera are allowed to remain. Washing, if done at all, is very sketchy. As a result, deterioration is often more rapid than with round fish, since the flesh is cut open, the bacteria laden slime can enter, and the blood and bits of viscera are excellent culture media for spoilage organisms. The fish must be thoroughly cleaned and well washed in clean water, if this operation is to delay, not promote spoilage.

In smaller fishing boats the cod should not be exposed directly to the sun and air, even if the trip to shore is only a few hours. A covering of some sort, such as a tarpaulin should be stretched over the fish, but should not rest directly on them. Space must be left for the circulation of air. Several buckets of clean sea water should be thrown on the fish to rinse off as much slime as possible when they are taken into the boat.

ATLANTIC COAST METHODS

On the Atlantic coast, most of the fish salted in the New England industry are surplus stock from the fresh-fish market. If the demand for fresh fish is light, these fish are sold principally to the fish-curing firms in Gloucester, Mass., or Portland, Maine. The vessel fishery where the catch was salted aboard ship has almost disappeared. Some shore fishermen along the coast of Maine still operate but the fish are cured ashore. The vessel-fishery cure is still carried on in the Maritime provinces of Canada, and in Newfoundland, though even here it is somewhat reduced in size.

Butchering and Splitting

In New England when the fishing vessel is docked the fish are taken out of the crushed ice and pitched into baskets of buckets holding about 2 bushels. These are hoisted out of the hold and emptied on the wharf where the cod are graded into three sizes, large, medium, and small (also called "snappers") and "scrod"). They are then pitched into boxes on platform scales and are weighed. All of this pitching is done by the single tine forks of pews discussed under handling. After weighing, the fish are forked into two wheeled carts, taken to the splitting tables, usually located in a nearby fish house, and dumped on the floor.

Then the cod are forked into racks near the end of the splitting table. Here a workman known as a "throater" grasps the fish, holding the head with his left hand, with the belly side of the fish up. He makes a cut through the isthmus and cuts down around the backbone following the curving line just below the gills. The head is held over the edge of the table and is broken off at the first vertebra with a quick snap, leaving a more or less ragged top. If the fish has not been already cleaned on the fishing vessel it is cut down the belly to the vent, and passed to a "gutter" who removes the viscera. The livers and roe are usually saved.

The cod then go the the splitters at the other end of the table. The splitter as a rule, uses a special knife, rounded at the tip of the blade, and with the blade curved slightly, flatwise. With the back of each fish

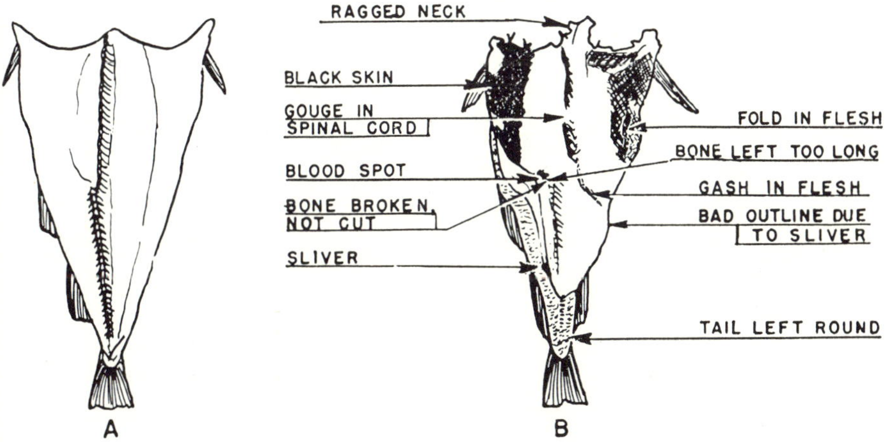

Figure 4.—Splitting cod for salting: (A) Good splitting; (B) Poor splitting.

braced against a cleat nailed on the splitting board, the cut made previously for gutting, from the throat to the vent, is continued along the left side of the backbone to the tail, making the cut no deeper than necessary for cutting out the backbone. A horizontal cut is made through the backbone about two-fifths of the distance from the tail and it is loosened so that the splitter can catch the lower end in his fingers. Grasping this with his left hand, he cuts under the backbone toward the head of the fish and separates the forward three-fifths of the backbone from the body, the lower two- fifths remaining. The edge of the knife blade should be held close against the backbone. If this is not done, much flesh adheres to the backbone and is lost. A short, shallow cut is usually made under the edge of the remaining section of the backbone, to permit the removal of blood spots and obtain better penetration of salt. When finished the cod should be split evenly along the backbone from head to tail, with the backbone cut, not broken, three fifths of the way down, with no round tail and no ragged edges, no sliver and no gashes. The difference between good and poor splitting is shown in figure 4.

Washing and Cleaning

The split fish are thrown into tanks or vats of clean sea water where they are washed after soaking diffused blood out of the flesh. Sea water is preferred since fresh water tends to soften the fish. Washing is important in determining the quality of the product, and in inhibiting decomposition during curing. Unfortunately, washing and cleaning are often superficial and careless. Careful thorough cleaning is essential to a well-cured product. Where running water is not available the standing water should be renewed frequently. Debris and offal accumulate quickly in standing water so that unless the water is changed every few minutes, washing becomes only a process of contamination. A small scrubbing brush should be used on the skin side, and all slime removed, paying particular attention to the area of the back fins. McPherson (1935) states that:

> "A brush should never be used for the face of the fish for rough surfaces should be avoided and this can be done with a cloth and plenty of water. The fish should be washed free of all blood and all pieces of gut, in fact it should be in as clean a condition as it would be if it were being prepared for cooking; indeed it is more important that it should be thoroughly clean before preserving than before cooking for the extent of cleanliness affects the preservation more than it would the cooking and eating."

The black lining of the belly cavity known as "black skin" in New England, scrubs off easily in washing. Although removal of the black skin is not essential to preservation the appearance is improved, and for this reason it must be removed if the fish are being prepared for the domestic trade as boneless or for any of the packaged preparations of salt cod. After washing, the fish are loaded on two-wheeled carts to drain while being hauled to the salter.

Salting

Salting is done in either of two ways, the pickle cure and kench cure. Most of the New England fish are pickle cured, though in winter they may be kench cured. Most of the fish salted ashore in the Maritime Provinces of Canada are pickle cured but all of the fish salted aboard the fishing vessels are kench cured. Some of the shore caught fish in Newfoundland are pickle cured, but much is kench cured.

Pickle Cure

Pickle cured fish are salted in tanks or vats of wood or concrete. Butts or hogsheads, usually old molasses barrels, are often used. These butts are about 3 feet in diameter and four feet high. The salters sprinkle a thin layer of salt on the bottom, then place a layer of cod, flesh side up, with the layer level and the fish overlapping as little as possible. Good salting requires skill and experience. The fish well spoil if the amount of salt is insufficient, while an excess results in inferior flavor and undesirable texture. Small or thin fish require less salt than large thick ones and less salt is necessary in cold weather than in warm. As a rule an even layer of salt thoroughly covering the fish and leaving no uncovered places or finger marks is sufficient. At one time Lisbon (a solar salt from Portugal) or Cadiz was used. Later Turks Island (a salt from the West Indies) was favored. Now, however, most New England cures use a domestic mined salt as it is better grade and standard in quality. While fisheries authorities have urged the use of salt by weight rather than measure, this is not often done. The amount of salt used in the domestic pickle cure averages about 17 pounds of salt to 100 pounds of split fish.

The tanks or butts are packed in alternate layers of fish and salt until the level of the rim of the container is reached. The fish are packed about 18 inches to 2 feet above the rim but the last few layers are packed skin side up and a heavier layer of salt is spread on top. As the salt extracts moisture the fish settle slowly and within a day or two sink below the level of the tank rim. It is often necessary to add an additional layer or two of fish at this time. No brine is added, as the extraction of moisture by salt should form sufficient brine. This brine or pickle tests 87° to 97° or an average of 90° salinometer. It should not be allowed to weaken below 90°.

When the fish has settled, salt is heaped on top not only to keep the cod under the surface of the brine but to make sure the pickle at the top remains fully saturated. The fish are usually "struck" through at the end of about 12 days but are allowed to remain at least 3 weeks. Pickle cured fish can safely be left in the tanks up to a year if the fish are held below the surface of the brine and it is kept saturated (figure 5).

Kench Cure

The kench cure is used when the cod are salted on board the fishing vessel, or in cool weather ashore. In kenching, a layer of salt is spread on a suitable flat surface, such as the deck of a vessel's hold. There must be good drainage. A layer of fish is then piled flesh side up, in a double row, with tails in and napes out. A thick layer of salt is spread over the layer of fish. The piling of alternate layers of fish and salt is continued to a height of about 4 feet. Two points should be kept in mind in kench salting (1) to make the center higher than the edges so that brine will drain outward, and (2) to use more salt on thick fish than on thin ones and with the most salt spread over the thick part of each

Figure 5.—Salting cod in butts.

fish. The fish should also be spread out as flat as possible so that there are no kinks or folds. The top layer of fish is spread skin up (figure 6).

The amount of salt used in the kench cure varies with the length of time the fish are to remain in the salt, the type of product desired, weather conditions, and the type of salt used. In the domestic industry, using a good grade of solar salt or mined salt, the amount will average about 20 pounds of salt per 100 pounds of fish, though as much as 30 pounds of salt may be used. In Canada, for the kench curing of heavy-salted fish, 100 pounds of salt will be used for about 350 pounds of split fish. In the domestic industry the fish are considered struck in from 14 to 21 days. In Canada the length of time is about the same. In Newfoundland it varies from 10 days to 4 weeks. McPherson (1935) gives 10 weeks as the outside limit, with the heaviest salting, that cod should remain in the kench. Kench-cured fish cannot be left long if the temperature is 60° F. or over, as they soon start to spoil. Kenched fish are also much more liable to discoloration.

Drying

When the fish are to be dried they are taken from the kench or pickle and washed thoroughly in clean salt water or in brine to remove any slime, dirt, or excess salt. They are piled in kenches (stacks) with

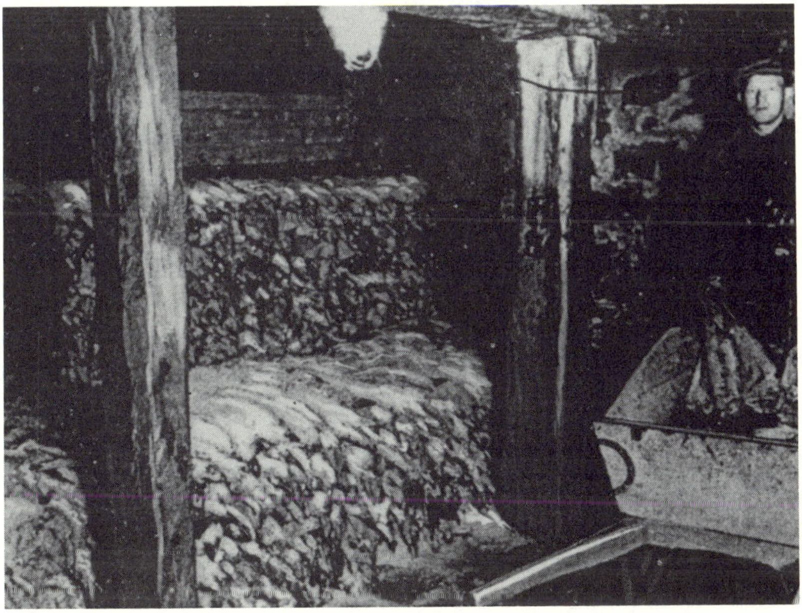

Figure 6.—Salting cod, kench cure.

the flesh side down, except for the first two layers, on slatted wooden racks from 3 to 9 inches above the floor. These kenches are about 3 feet high. The purpose of this procedure, which is known as "water horsing", is to press out some of the moisture and give the fish a smooth surface. The kenches are often weighted down to press out additional moisture. After 24 hours the fish are often re-piled, but with the layers reversed, that is, with the bottom layer on top. If the weather is unfavorable or if the drying flakes are full, the fish may remain in water-horse for as much as 10 days or 2 weeks. The kenches are torn down and re-piled at intervals as this aids in the drying action and helps to prevent spoilage. If the weather is good and the flakes are nearly empty, 48 to 72 hours water-horsing should be sufficient. The fish are now in the stage known as "green-salted."

Air Drying

The traditional method of drying of the Atlantic coast is in the open air on "flakes." This method is still followed generally in New England and Newfoundland. Improved types of artificial driers have almost entirely replaced open-air drying in the extensive cod-curing industry of

Figure 7.—Drying salt codfish on flakes.

Nova Scotia. The Pacific cod fishery of the United States has always depended on artificial driers.

In the New England industry, flakes are latticed frames standing about 3 feet above the floor. They average 8 feet in width and as long as convenient to the location of the drying yard. The bed of the frame is made of triangular wooden strips with a one-inch base, fixed about 3 inches apart. The fish rest on the sharp edges about 4 inches apart, an arrangement whereby the maximum circulation of air about the fish is obtained. The old style of brush flake still used in Newfoundland has disappeared from New England. The flake yard must be located convenient to the saltery. Flakes are often constructed on the roof of the saltery building itself.

If the fish are carefully dried under favorable conditions a much lighter colored product is obtained. Therefore, careful workmanship in drying is essential to prepare a product of best quality. On hot, bright days a canvas awning is stretched about 3 feet above the flake so that the fish will be sheltered from the direct rays of the sun, which cause yellow discoloration or "sunburn" especially with freshly spread fish. The fish must also be protected against dampness.

When there are signs of wet weather, the fish are gathered in small heaps and covered with flake boxes. These are rectangular wooden boxes with a peaked roof, made of three- quarter inch rough boards and are about 22 inches wide, 38 inches long and 14 inches high. A typical flake yard in New England is shown in figure 7.

The amount of drying depends on the grade of fish and the market for which it is intended. Full-pickle fish is dried the least especially if it is to be made into codfish bricks and middles. For the best domestic trade the product must not be too dry. It should retain a moisture content of from 60 to 65 percent. Also salt cod does not skin readily if too dry./ Therefore, in favorable weather it is given one drying only. This means about 10 hours of warm sun and good breeze.

The rate of drying depends on air conditions. According to New England curers the temperature of the air should be about 70° to 75° F. to obtain the best results. The wind velocity should be at least three miles per hour and may be higher, and the air should be less than 70 percent saturated with moisture. The percentage saturation of air at any given temperature is usually called relative humidity. Ideal conditions of air temperature, velocity, and humidity are found only during a part of the year, most often in spring and autumn.

Artificial Driers

Optimum drying conditions may be obtained at any season of the year without regard to weather conditions if an artificial drier is used. Artificial driers have been in use for many years on the Pacific coast and tried experimentally in New England. In Atlantic coast areas of the

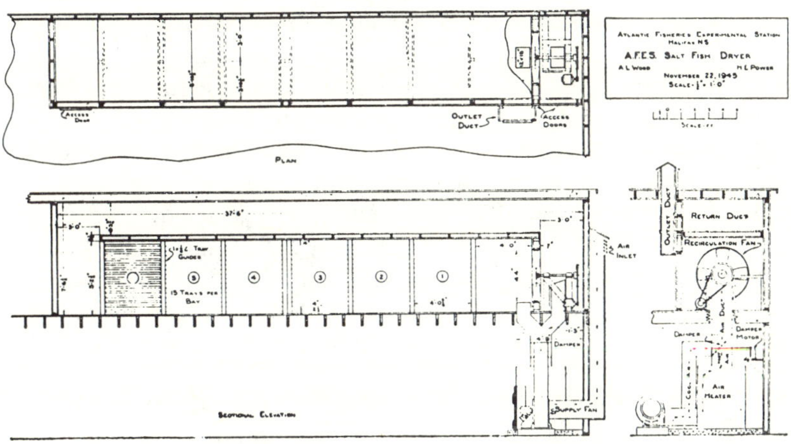

Figure 8.—Plan of artificial drier developed by the Atlantic Experimental Station, Halifax, Nova Scotia.

United States they are, however, used commercially only for special products. In the past there have been objections to the use of the artificial drier in New England. It has been claimed that the fish become case-hardened and do not dry so well inside, that they are more brittle, and that the color is not so good. It is also asserted that fish "scald" easily in these driers. Control of temperature and relative humidity have been the biggest obstacle. McPherson (1935) developed an improved drier but it could be used only with fish already partially air dried. The most extensive studies on the artificial drying of salt cod have been made at the Halifax laboratory of the Fisheries Research Board of Canada. As a result of this work nearly all fish curers in Nova Scotia are using the drier designed at the Halifax laboratory. It is also in use in the remaining provinces of the Maritime in eastern Canada.

For the successful artificial drying of salt fish, it has been found by the staff of the Halifax laboratory, that the minimum air velocity over the fish should be about 200 to 300 feet a minute. Lower velocities decrease the rate of drying. Higher velocities do not increase the rate and only add to the power cost. The optimum air temperature has been found to be 75° F. Higher temperatures than this may cook the fish, while at lower temperatures the drying is too slow. The recommended relative humidity of the air in the drier is from 40 to 50 percent. If the relative humidity does not fall within these limits the rate of drying is slower and the surface of the salt fish is rough, making the fish lower in quality. The investigators at the Halifax Station report that a major proportion of the cost of continuous drying consists of the construction and operation of the dehumidification equipment necessary to maintain relative humidity of 40 to 50 percent during the summer. When the drier is operated only during the cooler months, the cost is relatively low and construction and operation are comparatively simple.

The drier (figure 8) is of the tunnel type, built of wood with a wallboard lining. It is 28 feet long, 5 feet high, and is divided into 4 bays. Each bay contains 10 trays, made with a wooden frame and covered with chicken wire. A tray is 4 by 5 feet and holds 40 to 50 pounds of boneless or green salt-cod. A return air duct constructed on top of the drier enables the air to be recirculated. As a rule, one bay is loaded or unloaded at a time. According to Linton and Wood (1945):

> "The supply fan blows the outdoor air either through the air heater or directly into the drier through a bypass. A mixing damper set in the inlet duct adjusts the relative volumes of incoming warm and cold air to give the desired temperature. The position of the mixing damper is control-

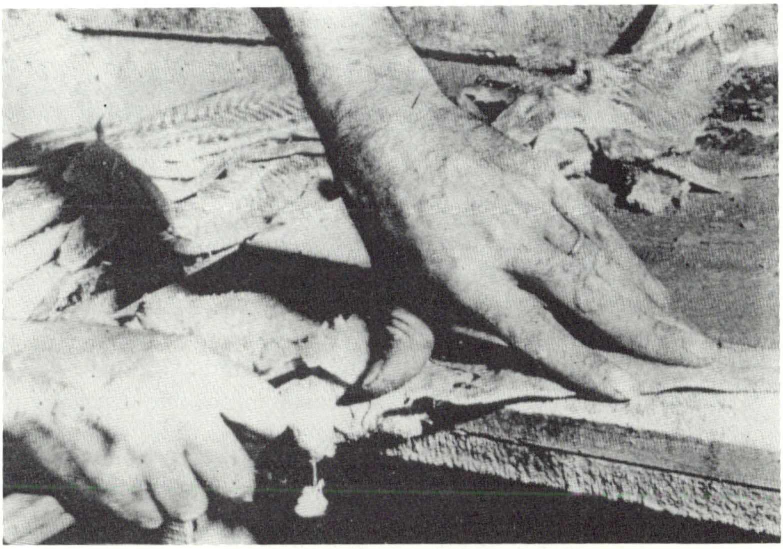

Figure 9.—Splitting salt codfish for bricks.

led by a damper motor and thermostat. If the temperature of the drier falls below 75° F., the mixing damper closes the bypass and admits more warm air from the air heater. Conversely, As the temperature rises the mixing damper closes the warm air duct from the heater and opens the bypass to admit cool outdoor air. The supply fan admits about 1,500 cubic feet of air per minute and requires a one-third HP motor. The air heater in the present drier is a warm-air furnace, burning coal or wood. Where steam is available, the usual finned steam heater may be used. The obvious advantages of the warm-air furnace are low initial cost, ease of installation, and avoidance of danger from freezing when operated intermittently during the cold weather. The average air velocity over the fish of 250 feet per minute is maintained in the drier by a 42 inch diameter recirculating fan. This propeller fan handles about 6500 cubic feet of air a minute and is operated by a one-half horsepower motor. The entering warm air is mixed thoroughly with air from the return duct by the propeller fan and the mixture is then distributed uniformly over the fish. Waste air is exhausted to the room at the end of the funnel. The return duct carries about 5,500 cubic feet of air a minute and is large enough to avoid appreciable resistance to air flow.

The relative humidity in the drier depends mainly on the volume of dry outside air drawn in by the supply fan. A hair humidstat, that is, an instrument sensitive to changes in the relative humidity, is installed in the drier and starts and stops the supply fan automatically as the relative humidity increases or decreases. If the relative humidity in the drier falls below the desired 40 percent owing to the intake of dry outdoor air, the supply fan stops automatically and as the relative humidity increases owing to evaporation of water from the fish the supply fan starts again. In damp weather the supply fan runs continuously. However, the installation of an ice box for cooling and drying the incoming air enables the dryer to be operated."

Water-horsed fish when thoroughly struck contain from 58 to 59 percent water. This moisture content must be reduced in the drier to about 53 percent for one-pound packaged, boneless fish and to about 48 to 50 percent for boxed, boneless strips or middles. Salt cod dried for export should have a moisture content varying from 35 to 43 percent depending on the market for which it is prepared. In the drier just described boneless salt- cod fillets require from 7 to 15 hours to dry depending on the size of the fillet and the moisture content desired. Export or hard-dry fish take about 40 hours. This drier will hold from

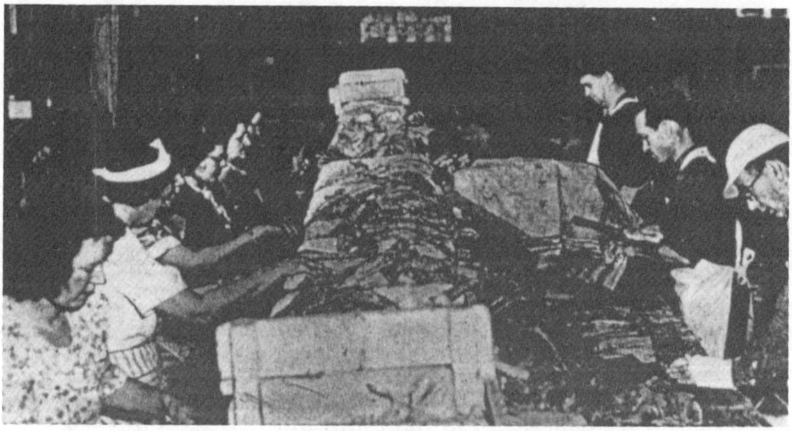

Figure 10.—Preparing codfish bricks.

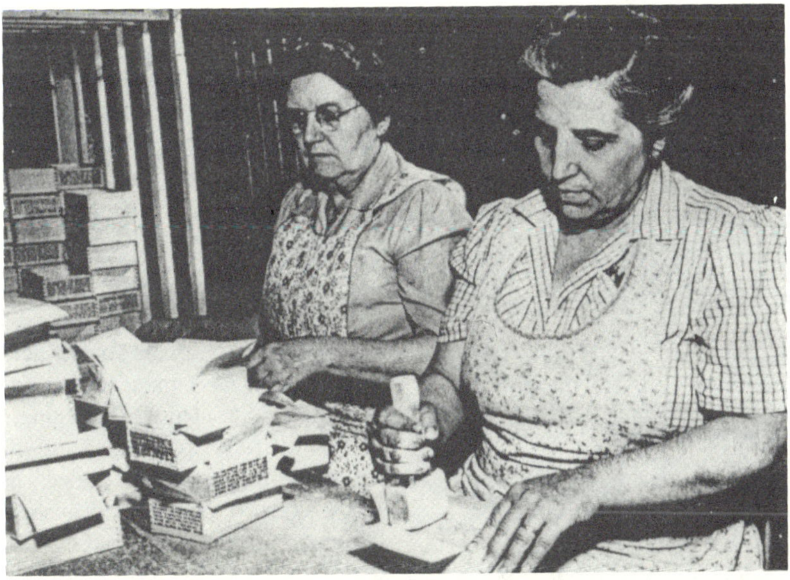

Figure 11.—Handpacking 1-pound wooden cartons of skinless and boneless cod.

1,500 to 2,000 pounds of green salt fish and thus has a capacity of about 1500 pounds of boneless or 350 pounds of hard-dried cod in a 9-hour day.

PACKAGING SALT COD

A considerable amount of the salt cod now packaged in the United States is cured in Canada, then imported as green salt fish into the United States, where it is dried, skinned, and boned.

Skinning and Boning

Fish intended for the better grade retail trade in the United States are skinned and boned. From the warehouse or flake yard the fish are taken to the "skinning loft" where they are first graded for quality and size. The best quality cod are thick, of a clear, uniformly white color, no blood stains or gashes, a smooth surface, and have a "sweet" odor. In warm weather or if the fish must remain in the skinning loft for some

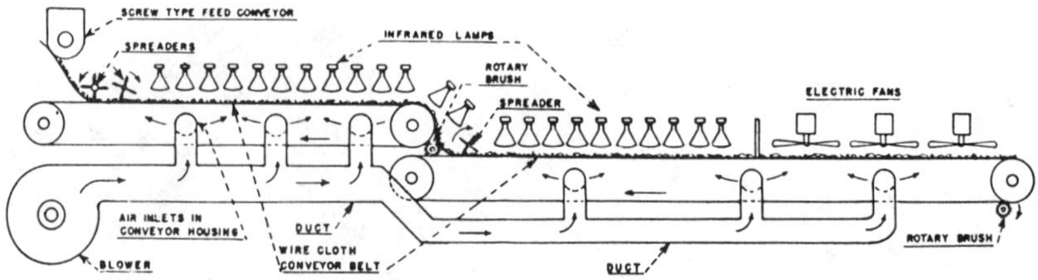

Figure 12.—Diagrammatic sketch of an artificial drier for shredded cod.

time, they are dusted with fine salt containing 0.4 percent boric acid. This is used as a preservative and as an inhibitant against reddening.

Figure 13.—Packing salt cod (export cure) in drums, Canada.

The first step in skinning and boning is to remove the dorsal and ventral fins. The workman starts skinning at the nape, pulling the skin loose towards the middle of the back, after which it is pulled back down to the tail. If the fish has been properly cured, the skin can be stripped clean without tearing the flesh. The nape bone is removed with a small iron hook known as a "bone hooker". This implement is about 8 inches long with a curved shank and sharp point. The tail is cut off, after which the fish is turned over, cuts are made under the edge of the section of the backbone still embedded in the flesh, so that it can be removed, and the pectoral fins are cut off. Formerly, these fish were called boneless, but under a ruling of the Food and Drug Administration the fish must be entirely bone free so they are passed to other workers who remove the small rib bones with pliers. The fish are then trimmed, cutting away any stained, discolored, or ragged sections, and are ready to be made into codfish bricks or the various styles of packages distributed to the trade.

Making Codfish Bricks

The method of making codfish bricks described by Stevenson (1899), Cobb (1926), and others, is still used in the smaller plants and in making bricks from hake or pollock. The larger packers have made some changes in the preparation of codfish bricks, speeding up production. This process of packaging codfish bricks in a large United States plant is briefly as follows: Skinned and boned fish of good quality are carried to a cutting table which forms the opposite side of the packing table. Here they are cut into pieces of the proper length. Several fish can be laid on top of each other and cut at one time. They must be placed carefully, however, so as to secure the best cuts from the thick parts along the back. These thick pieces are sometimes split in two to make good outside pieces for the brick (figs. 9 and 10).

After cutting, the pieces are sorted roughly and placed in small bins running down the middle of the packing table. A conveyer belt brings empty molds down the packing table. Each mold is of a size to hold one pound of fish. Girls working at the packing table take the empty molds. One large piece of fish is selected to serve as the bottom and one side, and is placed in position. The center is filled in with small strips. Then another piece is packed in acting as the top and other side.

The filled mold then travels down another conveyer belt to a revolving turret-type mechanical compressor. Here the fish in the mold is compressed to such an extent that it can be removed from the mod as a solid brick. The bricks travel from the pressing machine along a conveyer belt to a wrapping machine. En route women remove the molds and inspect the bricks for imperfections. A covering of vegetable parchment paper is placed on the bricks in the wrapping machine from which they pass to a packaging machine where they are placed in lithographed pasteboard cartons. The cartons are then packed in cases, usually 24 one-pound bricks per case.

The older method is almost entirely an individual hand-pack operation. The most usual of the older types of presses for making codfish bricks consists of a sliding metal box having two or three compartment molds 6 inches long by 3 inches wide by 3 1/2 inches deep. This will take a two-pound brick. The length and width are the same for one pound bricks, but the depth is 1 3/4 inches. When the mold is slid into position a die operated by hand or foot pressure is forced down into the mold, compressing the fish. The molds have either two or four slots in the sides. Cotton strings are run through the slots so as to pass around and under the block of fish. The strings are placed in position in the empty mold after which the brick is packed. The method of packing is as described in the preceding paragraph, except that the pieces for a brick are all selected and weighed before packing. When a mold is filled it is slid into place under the die. The brick is then tightly compressed for a few minutes. While it is under pressure the number two mold is packed. The strings are tied on number one, the pressure is released and the brick is removed. Number two is then placed under pressure while number one mold is being packed, and the brick maker proceeds as before. The finished package is then wrapped by hand in vegetable parchment paper and placed in an individual lithographed pasteboard carton.

Bricks are also hand packed in wooden cartons holding one, two, or three pounds. Selected fish is packed as described in the wooden dovetailed boxes previously lined with vegetable parchment or waxed paper. The fish is then pressed down with a hand tamper (figure 11), the paper is folded over the top and the cover is nailed on.

Shredded Cod

Another packaged salt cod product is "shredded cod", also sometimes known as "fibred codfish" and flaked codfish. The salt fish used for its manufacture is usually whole fish of the same grade as is used in making bricks, fish of good quality, but too small to use in bricks. The fish are dried somewhat more than for boneless fish and all bones are removed. A quintal (112 lb.) of pickle-cured fish will make about 60 pounds of fish ready for shredding.

The pieces of fish are first run through a shredding machine which tears the flesh into small, fibrous bundles. The moisture content must now be reduced from approximately 60 to 40 percent in an artificial drier. Several types of these driers are in use, mostly cabinet or tunnel driers of varieties used for many years in the drying of vegetable foods. Most of these are batch operations, have low capacity, are more or less difficult to control, and are expensive to operate. One large fish curing firm has developed a continuous drier designed to overcome these objections (Burton 1943).

The drier is of the tunnel type. The shredded fish enter the upper section of the drier through a screw-feed conveyer, falling on a conveyer belt of stainless steel wire cloth. The fish first pass under two sets of revolving paddles which spread the shredded material over the belt. The conveyer belt then passes under a battery of infra-red lamps which heat the shredded codfish to about 150° F. Air is blown through the material from ducts below the conveyer belt driving off the moisture extracted by the infra-red lamps, and preventing over-heating. After passing the first battery of lamps the fish drops to a second conveyer belt where it is spread out by revolving paddles (fig. 12). The operation of the first stage is repeated. After passing under the second group of lamps the shredded fish passes under three electric fans installed for cooling effect and to remove additional moisture. The dried material drops from the end of the belt into boxes in which it is carried to the packing table. It is hand-filled into wax-paper-lined pasteboard containers to a net weight of 4 ounces. This drier has a capacity of 4,000 to 4,300 pounds a 7 1/2 hour operating day.

Shredded cod is also packed in 5 and 7-ounce glass tumblers. The fish is first sifted to remove stray particles of bone, then filled into the containers either by hand or machine. The jars are fitted with metal tops and sealed in a vacuum closing machine. This product is not heat sterilized but due to the low moisture and high salt content it should keep for about two years. The glass tumblers should be kept from strong light to avoid oxidation which causes loss of flavor. Shredded fish is used principally for making codfish cakes.

Strips and Middles

Skinned and boned cod are also packed as "strips" and "middles". Strips consist of one-half the fish cut down the middle. The napes, tail, and edges are cut off leaving an even thick piece of salt fish. This not only improves the appearance but fish may be packed more solidly in the shipping container. Strips are packed in 10, 20, and 40 pound wooden boxes.

Middles consist of the whole fish with the napes, tail, and thin part of the belly flesh trimmed away leaving one large piece of thick flesh. Middles are packed in boxes of the same size as strips. In the same class is skinned cod in 100 pound cases. This pack consists of the finest whole cured cod, skinned and boned. It is graded according to the number in the case, usually as "large whole" or "extra large whole."

Hard Dry or Export Cod

Cod intended to be marketed in the southern states or for export is dried to a lower moisture content than for the domestic trade. Either pickle or kench-cured fish may be used. A larger percentage of the moisture is removed with salt in kench curing, and the fish are reported not to make so good a dried product. The fish are taken from pickling tanks or kenches, washed in sea water, water-horsed, and set out to dry on flakes as described previously. These are dried for 3 or 4 days, then taken in and piled in kenches to "sweat". Moisture is brought to the surface from the inside largely by pressing. The fish remain in the kench for 2 or 3 days after which they are again spread on the flakes to dry for a day or two. The cod are then repiled in kenches and sweated for 3 or 4 days. After a third drying of 2 to 5 days they should be ready for packing. This fish product is known as "hard dried" or "export cure." Some use of artificial driers is made, particularly at seasons of the year unfavorable to air drying. No accurate determinations of moisture content are made but the appearance and texture are noted carefully from time to time. The export fish is usually ready for packing when dried sufficiently hard to withstand the pressure of the thumb in the thick part of the flesh without retaining the impression and there is no visible moisture in the fin folds.

Hard dried fish are packed in drums, boxes, bales, or bundles as the order may require. The regular drums, the package mostly used for export, have net contents of 50, 100, 200, 300, and 448 pounds. The 448 pound drum is the most popular size; 100 pound boxes are also coming into demand. When packing in drums, several layers of fish are carefully arranged on the bottom in circular fashion with tails to the center and flesh side up. Then a single layer is packed skin side up. The cod are than tamped thoroughly with a heavy wooden tamper. Several more layers flesh side up and a single layer skin side up are packed in and the tamping is repeated. The alternate packing and tamping is continued until the drum is filled. It is necessary to pack several inches above the top of the drum to obtain the required net weight, and a jack screw or hydraulic press is required to press the fish down sufficiently for the head of the container to be put in place (figure 13).

LIGHT CURES

Slack Salted Cod and Pollock

Some of the shore fishermen of the Maine coast prepare small quantities of salted cod and pollock for local use, mostly in October and November. This product is known as slack salted. The fish are caught near shore and are headed, eviscerated, slit, and washed on shore. They are "corned," that is, very lightly salted by being packed in salt from 12 to about 48 hours. The fish are then rinsed, and hung from over-head beams in a fish house or warehouse where there is good cross ventilation. They are left here until rather hard and dry. While this product is lightly salted it is dried even more than for the export trade. The flavor is stated to be superior to the ordinary commercial grades. It is sold entirely in the coastal towns of New England.

Gaspé Cure

The best known of the mild cures is the Gaspé cure prepared mostly on the Gaspé Peninsula of Quebec, but also at Caraquet, in New Brunswick, on the Magdalen Islands, and in certain parts of Newfoundland. It is considered to be a high-class product and at one time had a good market in southern Europe, especially in Italy. Some Gaspé cure fish are imported into the United States for sale to the Italian-American trade and the French-Canadian trade of New England.

PACIFIC COAST COD INDUSTRY

Cod only are cured on the Pacific coast. Haddock, hake, cusk, and pollock are not present. The Pacific cod banks have never been fully explored. The known areas are larger than the Grand Banks of Newfoundland, but there are extensive sections where occasional catches of cod have been reported which have never been thoroughly explored or fished commercially. While cod are found in the eastern Pacific from Oregon northwards they are taken for curing only off the coast of Alaska.

As a result of the enormous demand for protein food during World War I there was a boom in the Pacific cod fishery, and the catch rose to about 30,000,000 pounds annually from 1914 to 1918. In the postwar depression of 1921-22, it declined to 9,000,000 pounds, gradually increased to 24,000,000 pounds in 1926, decreased to 16,000,000 pounds in 1927, remained at approximately that level for the next 10 years. The California fleet abandoned the Alaska cod fishery in 1938 and then it declined to a level of 10,000,000 pounds, where it remained until 1942. In that year, the cod-fishing grounds became a war theater, most of the vessels were taken for war purposes, and only one vessel was able to fish. There was no cod curing in the Pacific in 1943, other than the very minor operation of a few shore stations. A single vessel operated in 1944 and 1945. Salt codfish is being shipped from the Atlantic coast in carload lots of whole fish, generally to firms which formerly packaged and merchandised Pacific coast salt cod. The postwar status of the Pacific salt cod fishery depends on the anticipated greater exploitation of fisheries in western Alaska by shore stations and factory ships. At present, a once flourishing fish-curing industry has almost disappeared—a war baby of World War I, a casualty of World War II.

The methods of the Pacific cod fishery and the salt-cod products prepared, generally resemble those of the Atlantic coast, but there are a few important points of difference. In the first place, in the Pacific the hand line is used almost entirely, to the exclusion of other forms of gear. Secondly, practically all of the cod are cleaned and salted either on board the ship or at shore salting stations in Alaska and are later transported to the home station in the State of Washington, where the fish are dried and packaged for distribution. The vessel-caught cod are not iced, as are those caught on the banks of the Atlantic coast, but are salted in kenches in the hold of the ship. Those caught by the fishermen near the shore stations are salted in tanks at the station but are later shipped to the states for drying. The drying is done principally in artificial driers. The amount of salt used in curing is appreciably greater on the Pacific coast. As a rule from 25 to 30 pounds of salt per 100 pounds of fish are used in the tank or pickle cure. In the kench cure of the vessel fishery the amount of salt averages from 35 to 40 pounds per hundred. The history, statistics, and methods of the Pacific industry are given completely by Cobb (1926). His publication should be consulted for additional detail.

Chapter Eight

Drying Salt Cod: Foreign Cures

FRENCH METHOD OF CURING COD

Bordeaux is the center of the French cod-curing industry. Some of the salt cod used in France is cured on the islands of St. Pierre and Miquelon. The method is much like that used in Newfoundland. Much of the catch, however, is salted on board the fishing vessels on the Grand Banks and brought to France to be Cured. The fish kenched in the hold of the fishing vessel are transferred to warehouses ashore where they are held in bulk with some additional salt between the layers of fish. The cure is completed as orders for dried fish are received.

Splitting and Salting

In splitting, the French allow more of the backbone top remain in the fish than other curers. Therefore, to remove the blood under the section of backbone left in the cod they use a long-handled iron spoon as a blood scraper. The fish are always washed thoroughly in clean sea water after splitting and before being put in salt. The amount of salt

Figure 14.—Air-drying salt cod; French cure.

used is reported to average one kilogram (2.2 pounds) to 2 kilograms (4.4 pounds) of fish. Dieuzeide and Novella (1942) give the amount of salt as one kilogram of salt per kilogram of fresh fish. This includes salt used for all purposes, including the preservation of bait and for repacking. A solar salt manufactured on the French Mediterranean coast is used.

Construction of Drying Racks

Instead of air drying with flakes or on rock covered beaches, the Bordeaux curers use vertical racks from which fish are hung by the tail. This type of air drier is practical because it takes up so little space, the cure is finished in a short time, varying from two to six days, the labor used is reduced to a minimum, and the product is of excellent quality, although it does not keep so well as that cured by some other methods.

The racks are constructed as follows: A number of posts are driven into the ground vertically in a straight line from west to east and about 4 1/2 feet apart. Two rows of laths are nailed along this line of posts, just far enough apart to allow the tail of a codfish to be pushed between them. Small cleats of wood are nailed across each pair of laths to keep them from springing apart; the laths are 3/8 to 5/8 inches thick and 1 1/2 to 2 inches wide. The construction of these drying racks may vary slightly. Some are built in square compartments with laths fastened alternately on the north and south sides of the posts and with an alleyway about 2 feet wide between each row of compartments. On other racks, the laths are all fastened on the north side of the posts and each row of racks is set 3 yards apart, in order to prevent the shade from the row in front reaching the one behind (fig.14).

Drying

The cod are rinsed thoroughly and scrubbed to remove any traces of dirt or encrusted salt. After draining for a short time they are carried out to the racks. The tail of each cod is shoved in between the laths from the north side, with the back or skin side turned upward. The fish will bend down by its own weight and the face or flesh side will be turned toward the sun while the tail is jammed between the laths. When the fish are getting a little dry they will hang in this way even in a strong breeze. Some curers set a straw-colored thatched roof over their racks if the sun is at all hot. Others use no shade covering, but if they think that the sun is likely to burn the fish, the fish are twisted slightly so that the edges instead of the flesh sides are turned to the sun. In damp or rainy weather or when the sun is too hot, the fish must be taken down and stored in the warehouse. In the French codfish cure, the fish are not pressed at all during the drying. After hanging on the racks from 2 to 6 days, depending on the size and weather conditions, the fish are taken down and shipped to market.

While air-dried fish are preferred, artificial driers are used to some extent because of the uncertainty of natural drying conditions. Driers are of the tunnel type, 25 meters (82.5 ft.) in length, 3 meters in diameter (9.9 ft.) and 2 meters (6.6 ft.) in height. The cod are placed on movable racks, which are run into the drier. The cod are dried in a current of air at approximately 30° C. (86° F.). Length of drying time varies from 3 to 48 hours, depending on size and type of cure (Fillon 1929); Boury 1932, 1934).

ICELANDIC METHOD OF CURING COD

The Icelanders conduct most of their cod fishery in the open sea at a short distance from shore. The fishing grounds are close to the curing location, sometimes a distance of only a few hundred yards, with a maximum of about 60 miles. In no event are the fish more than 24 hours old when landed. Their fishing craft are principally motor cutters sometimes five tons or less, but more often are 80 to 100 tons in size. The fishing gear used consists of long lines, gill nets, and Danish seines. Fish taken by large trawling vessels using the otter trawl are occasionally used for salting, especially since the end of World War II.

Splitting and Washing

As soon as the fish are caught they are bled, usually by cutting the throat. When the cod are brought ashore they are gutted and split as soon as possible. The Icelanders split their fish differently from most other curers, in that the section of the backbone allowed to remain is on the right-hand side instead of the left. The backbone is cut slantwise over two joints instead of being broken straight across, which is thought to improve the appearance. The last 18 to 22 vertebrae remain, the number varying with the size of the fish. The Icelanders also split their fish very deep. After the fish are split they are washed thoroughly in large quantities of clean salt water and scrubbed with brushes to remove all black skin, blood, and other offal.

Salting

The salting is done in sheds where the fish are allowed to drain until all surplus moisture is removed. The drained cod are salted in kenches, using about 40 pounds of salt to 100 pounds of fish. The types of salt favored are: Torraveja, St. Pola, Cagliari, and Trapini. After the fish have been held in salt for 2 or 3 days they are resalted in new kenches, using very little salt this time, only about 1/8 of the former amount. They are left in the second kench for 4 to 6 days, when the fish are ready to be washed and dried, if the weather and season of the year are suitable. Fish caught too late in the fall to be "made" before the next year are salted so heavily in kenches that one fish does not touch the other. If properly washed and cleaned, and all blood removed, it is claimed that this fish will, in the following spring, be of about the same quality as if caught that season.

Drying

After the fish have been held the proper length of time in salt, they are washed thoroughly, and piled in small heaps until all surplus moisture has been removed. When the flesh feels slightly stiff the fish have been sufficiently drained. As a rule, 24 hours are required. If the weather is fair the fish are spread out to dry. If the day is unfavorable the cod are relaid, flesh side down, in square piles, from 100 to 150 fish in each pile. The fish are repiled each day as long as bad weather lasts. On the next good day they are again taken out to dry. The fish are spread on the rocks, flesh side up. Flakes or other types of drying platforms are not used.

Pressing

When the cod have had two good days of sun drying they are stacked in piles, and the cure continues much as in the Norwegian method, weighting the piles of fish more heavily as the cure progresses. When the fish are at the stage the Icelanders call "three parts dry" they are stacked in large piles, with about 7,000 pounds of fish in each pile. These piles are covered with mats or boards which form a roof, and are weighted down with enough rock to correspond to the weight of the fish in the pile. The cod are allowed to remain in these piles for 5 or 6 days, then spread out to dry for a day. At night they are gathered into big piles again and once more pressed down with a weight equal to that of the fish in the pile. This process is repeated until the fish are considered cured. If, because of unfavorable weather, the fish cannot be spread out after they have been put in the first large pressing pile, they are repiled every day and the same weight applied until they are considered cured.

The Icelanders use this heavy pressing in their codfish cure for these reasons: (1) The climate is damp and not very warm; as a rule the sun is seldom hot enough to burn the fish, although this may happen occasionally, and the weather is mostly cloudy or foggy; (2) their fish are heavily salted; (3) the fish are rich and thick, and stand a good deal of pressing.

Since the climate is chilly and damp, pressing is very important, and cure is accelerated by frequent pressing and repiling of the fish. The Iceland cod have taken away much of the Mediterranean market from Newfoundland and Nova Scotia because of good appearance and better quality. Though the fish are always a little pliable and are not cured so hard as Newfoundland and Norwegian fish, Iceland fish keep well in hot climates and are preferred to the hard-cured fish because they are less apt to be brittle and break. Fish cured in the early spring or fall when the climate is really chilly are, if sufficiently pressed and salted, superior to the hard-cured fish of the summer season and, even if a little pliable, will keep well in hot climates.

Figure 15.—Klipfisk spread out to dry in Norway.

NORWEGIAN METHOD OF CURING COD

The Norwegian dry-salted cod have acquired a good reputation for quality because of the care taken in curing and grading. They are known as "klipfisk" in Norway, and are so-called because they are dried on the rocks or "klippe" along the shore. The curing of klipfisk is an old established industry of great importance to Norway. Very large quantities are cured each year. Much more cod is prepared as klipfisk than stockfish. They are exported to France, Spain, Portugal, the Mediterranean, Central and South America, and the West Indies.

The largest production of the cod fishery is obtained on the northern coast of Norway, Lofoten being the center of the fishery. The fishing vessels sail north each year. The fish are salted on board, and later taken south to the drying areas. Some of the larger vessels sail to the Lofoten area, buy small lots of salted fish from the fishermen and sail south when they have obtained a cargo. The splitting, dressing, and salting are about the same as in the American vessel fishery; that is, they are cured as in our kench cure. The Norwegians are more careful in washing, however, and wash the fish once before and once after splitting. Cleaning is very thorough, especially in the removal of bits of coagulated blood or liver which might cause souring.

Salting and Washing

The fish are drained for about 2 hours after the second washing and before salting. The drained, split cod are salted in kenches about 3 feet in width, and the same in height. The salt used is generally Liverpool, Torraveja, Cadiz, Lisbon, Cagliari, or Trapani. The amount of salt used varies to some extent with the size and thickness of the fish, the type of salt, and the season. The fish are salted fairly heavily, but not so heavily as in the Iceland cure. The average is about 35 pounds of salt to 100 pounds of fish. The cod must remain in the kench from 25 to 30 days before they are ready for drying. The cod may remain in salt as much as 2 months.

The drying and pressing are usually done on the rocks in the vicinity of Christiansund and Aalesund on the "klippe." These are various stretches of smooth, white, clean rocks near the seashore (figure 15). When the fish are taken out of salt they are washed thoroughly in sea water. This is usually done at the beach nearest to the drying place. From 45 to 50 fish are thrown into the water at one time. No attempt is made to handle more than this number or the fish will be freshened too much, the sea water leaching salt from the flesh. In washing the fish, woolen mittens are worn. The fish are again washed carefully to remove excess salt on the surface, or other debris, and are brushed thoroughly to remove blood spots or pieces of black skin.

Drying

After washing, the fish are stacked in kenches about 18 inches in height. The pile is something like a low rick of cord wood. The bottom row of fish is laid skin side down with the other rows skin side up. The stack is usually covered with boards or canvas as protection against rain or sun. On the second or third day, the fish are restacked, reversing the order of the layers. On the third or fourth day, the fish are spread out on the rocks and given about 8 hours of air drying. No particular system is followed in spreading, the fish being simply laid flesh side up on the rocks, with no two fish touching. The first day of drying must be chosen carefully, for if the sun is too hot, a crust will form on the surface, preventing the removal of moisture from within, so that the fish will deteriorate.

After the first drying period the fish are stacked skin side up in round piles of 500 fish each. The diameter of a heap is about four feet. The piles are built highest in the middle so that the rain will run off. Care is taken to lay the fish out smoothly to prevent wrinkling. After two or three days of drying, the fish are usually stretched well, especially in the belly section, before being stacked, to remove all wrinkles and to give the fish a smooth appearance. This work is very important, not only to appearance but for longer preservation, because moisture always gathers in the wrinkles and is very difficult to remove entirely unless done in the early stages of the cure. Stretching is slow work but it is considered better to devote one day to it rather than to do it in a hurry in the evening when the fish are gathered in heaps for the night.

Pressing

When the fish have had two or three days of air drying they are stacked in round peaked heaps as before, except that the heaps are larger, with 1,000 fish in each pile. The diameter of each heap is still 4 feet, but the pile is higher. The fish are allowed to stand for 2 days, then given another day's drying on the rocks, after which they are repiled and held 2 days. The alternation of 2 days pressing followed by a day of drying is continued until the fish have had about 60 hours of good air drying. They should then be sufficiently dry. If there are spells during the curing period when the weather is not suitable for drying, the fish must be re-piled as often as possible, reversing the order every time they are re-piled. This is done because pressure extracts a certain amount of moisture, enabling the cure to continue, and to prevent the fish from becoming slimy.

Each pressing pile is covered with a small, round, peaked roof of wood about a foot larger in diameter than the heap of fish. This keeps off rain and acts as an added weight for pressing. Stones are placed on top of the roof to aid in pressing, but not to the same extent as used in Iceland. The best fish are said to be made by a long slow drying. Under favorable conditions the usual time required to dry klipfisk in Norway is about six weeks, but it may take as much as two months. When the fish withstand pressure of the thumb in the thick part of the flesh they are judged to be sufficiently dried.

BRITISH METHOD OF CURING COD

The commercial curing of cod in Great Britain is conducted largely in Scotland, where the most important centers are Aberdeen and two groups of islands, the Shetlands and Orkneys. Some curing is done in Grimsby and Hull, the two most important fishing ports in England. In Scotland the fish are usually landed fresh and split and cured on shore. In the deep sea fishery off Iceland at Grimsby and Hull part of the catch is split and salted on board fishing vessels, with curing completed on shore.

Splitting, Washing and Salting

The fish are headed and split as much in Newfoundland, Canada, and the United States, as is to be expected, since these fisheries were developed by fishermen from the British Isles. In Scotland, the backbone is usually cut through slantwise, over two joints of the vertebrae, as is done in Iceland. The backbone is cut off just below the anal opening. Mechanical equipment for heading and splitting the fish is in more general use than in North America.

The split fish are slid into a trough of fresh water. The blood is removed as completely as possible, black skin stripped away, and slime washed off. In larger plants, washing troughs are equipped with revolving stiff-bristle brushes, under a protective shield.

Shore-cured fish are usually salted in tanks or butts, the size and number depending on the scale of operation of the plant. According to Duthie (1911) a common size is 6 feet long by 4 1/2 feet wide by 3 feet deep with a few tanks 8 feet long for curing large fish. The fish are laid in the tanks flesh side up, alternating heads and tails to obtain an evenly packed layer. Care is taken to lay the fish flat to avoid folds and wrinkles which would interfere with curing and mar the appearance of the cured fish.

When butts or circular vats are used, the fish are laid in with tails to the center and napes to the side. Salt is scattered over each layer of fish. As in other cures, the amount of salt depends on the size and thickness of the fish, type of salt, and weather conditions. The amount of salt varies from 17 to 20 pounds per 100 pounds of split fish, 18 pounds being a general average. According to Duthie (1911) 45 pounds of salt is a fair average for preparing a hundredweight (112 lb.) of dry- salted fish from 2 1/4 hundredweight of gutted fish.

Duthie considers this amount heavy salting but it would be regarded as a light cure in North America as cod are only left in salt from four to six days. The fish are scrubbed in brine as they are taken out of the tank to remove any undissolved salt or other debris. The cod can be dried immediately but if this is not feasible, they are piled into heaps called steeples, with a sprinkling of salt over each layer.

Duthie (1911) states that when fish are cured aboard vessels—for instance at the deep-sea fishing off Iceland—brine tanks are not available and the fish have to be dry-salted. For this purpose "lockers" are fitted up in the sides of the vessels, and into these the fish are laid and salted in the same way as in tanks. More salt must be used in this case, however, and care taken to spread it liberally around the sides of the lockers, as well as over the fish, otherwise the exposed parts of the fish may be imperfectly cured. After lying for about a fortnight the fish should be turned over and resalted.

Drying

The oldest method of drying cod is on rock beaches, as in Norway. It is still followed to some extent in Scotland, mostly in the Shetlands. The fish are first laid out to dry, skin side down, but are turned frequently during the first few days of drying. At night the cod are gathered into small heaps, known as "clamps", with the skin side up. They are covered with a canvas or tarpaulin for protection against night dampness. As drying goes on these heaps are gradually made larger in size. After about two weeks the fish are gathered up into large heaps, covered with a tarpaulin, and weighted down with stones laid on planks. This is said to improve the appearance of fish and also to press out moisture. The fish are left for 10 to 14 days. This step is known as "sweating." At the end of the pressing or sweating period the fish are dried for about a week then piled and sweated again for about 5 days. After the second pressing the fish are dried for about 2 days, then stored. In good drying weather the entire drying period requires about 6 weeks. This is the same time reported for the Norwegian klipfisk cure. Flake drying is also used. It does not differ greatly from the method followed in other localities. Some use is made of portable wooden and wire flakes.

Kiln Drying

One drying method widely used is a very simple type of artificial drier. It is known as "kiln" drying and its advantages are that it requires less time than the air-drying method, and is not dependent on weather conditions. Almost any warehouse that is dry, reasonably fireproof, and well ventilated may be used for kiln drying, although special buildings are sometimes constructed. Duthie (1911) gives the following specifications for a typical drying shed. It is 60 feet long by 20 feet wide with side walls 8 feet high. Six heavy posts run down the center of the building to support the series of joists which run across the building from side to side at the top of the side walls. Over these joists, about a foot above the top of the side walls, four strong beams are laid, spaced evenly and running from gable end to gable end. Pulleys are fastened to these long beams spaced about 12 inches apart. The

Figure 16.—Kiln drying salt cod in Scotland.

"tenter frames" on which the fish are hung are suspended from these pulleys. Beams are fixed along each side of the building about 5 feet above the ground. Cleats for fastening the ropes connected with the pulleys are fastened to these beams. In the roof there are three pairs of hinged skylights arranged to open with pulleys and cords and in the walls below there are three pairs of windows. Cast-iron "hit-and-miss" ventilators, three on each side, are set in the side walls near the ground to regulate the draught. These four rows of oblong frames run down the building, the individual frame holding four rows of fish on each side. The fish, flesh side out, are fastened to hooks on the "tenter stick" and spaced about 5 inches apart, usually alternating heads and tails. When the frames are filled and hoisted, the lowest fish should be about 5 feet above the floor.

When the fish are taken out of the tanks or kenches where they have been salted, they are rinsed, then laid out to drain for a few hours before hanging on the frames. Drying is done with heat from portable coke grates having radiating iron plates covering the fire. Gas coke is used as fuel. The grates are filled and lighted outside, and the fire burns red before the grates are taken inside. This is to guard against discoloration by ash or smoke. The furnaces are set 12 to 15 feet apart and may be moved from time to time so that the fish will be dried evenly. When the fires are first brought in all the ventilators are open and the temperature regulated to about 60° F. Higher temperatures up to 80° F. are used when the fish are hard and in the final stage of drying (figure 16).

The fish are given an average of 48 hours of continuous drying over these grates. They are then taken down and stacked in heaps or "steeples" to sweat for two weeks. They are hung up to dry again over the coke fires. If the cod are of medium size, a second period of 48 hours should complete the drying. Large fish are built up into steeples again, weighted down with stones and planks. After a third drying period of 24 hours they are usually sufficiently dried for the export market. Duthie (1911) states that:

> "In the opinion of some experienced curers, the ideal cure can best be obtained by drying and sweating the fish outside until 24 hours 'kiln' drying will make them as firm as desired. The sulphury taint of the coke fires ... is scarcely discernible after only 24 hours in the drying shed, but it is claimed that even in this short time the fumes of the sulphur usually deter mites from attacking the fish afterward."

Chapter Nine

Spoilage in Dry Salt Cod and Other Cured Fish

No comprehensive study has been made on types and causes of spoilage in the various dried and salted fishery products. Research has been made from time to time on special problems, mostly in relation to the codfish industry, as cod and related ground fish have traditionally been the most important species in the fish curing industry. Available information is therefore based mostly on dry salt cod. Many of the problems are common to all species of dry-salted fish though there may be differences in detail, a fact which should be kept in mind in reading this section.

BACTERIAL SPOILAGE

Reddening

The most important cause of spoilage is reddening. Outbreaks of reddening are much more troublesome in some seasons than others and perhaps more attention is paid to the problem than to other types of spoilage. Some curers believe that reddening is of more or less recent origin, but references from the eighteenth century indicate that this trouble was known then. Reddening can be a source of considerable expense and annoyance to codfish packers. It is less common on the Pacific than the Atlantic coast. The comparative rarity on the Pacific coast is probably due to the cooler summer weather, and the use of a better grade of salt.

Reddening is a type of spoilage characterized by the change to a pink or very unappetizing red color on the surface of the fish. At first there is little change in odor and flavor, but both later make the fish unpalatable, as well as unattractive in appearance. This change may appear while the fish are being cured, when they are stacked in the warehouse, while they are being shipped, or after they have reached the market.

Reddening was one of the first problems to be investigated after the creation of the United States Fish Commission in 1871. It was also studied in Europe and Canada. The most valuable investigations contributing to a solution of the problem were made in 1911 by Bitting of the Bureau of Chemistry, U.S. Department of Agriculture; and in 1919 by Browne, an investigator of the U.S. Bureau of Fisheries. It was found that reddening is due to the growth of at least two organisms, the first a spirochete causing a pale pink color, and the second a bacillus resulting in a deep crimson color. Apparently these forms grow in close harmony; hence all shades of color from pale pink to deep crimson color are seen on salt fish. The Fisheries Research Board of Canada has devoted considerable attention to reddening in its studies on salt fish. Hess (1940, 1942) has obtained valuable data on the effect of environment upon the growth of red halophilic bacteria and on the effect of disinfectants and preservatives on these organisms. Gibbons (1935), Hampton (1938), Boury (1934), and others have also published studies on the subject. A review of the subject of reddening, with the more important citations of literature has been compiled by Tanner (1944).

Although there is confusion as to the nomenclature of the organisms involved in reddening, and it is believed that the same organisms have been given several different names, information is more definite as to the most favorable living conditions and possible methods of control. Growth of red halophilic organisms takes place in salt solutions containing between 5 and 15 percent of salt by weight. The temperatures necessary for growth range between 15° and 55° C. (59° and 131° F.). The range of salt tolerance can be increased in both directions from the optimum by gradual adaptation. The organisms grow well on heavily salted fish (unless dried to a low moisture content), brine, salt piles, and agar-containing fish stock saturated with salt. From the evidence at hand it is deduced that salt lagoons of the tropics are the probable source of most of the infections. Ordinary bacteria are usually destroyed after 10 minutes exposure to bright sunlight, but sunlight does not kill the reddening organisms because of their pigmentation. They have been found to have great resistance to ultraviolet light. Evidence to date points to sea salt used in fish curing as the major cause of reddening. Solar salts from both Europe and Africa have been found to be infected; mineral salts are apparently free.

Prevention of Reddening

Bitting (1911) gives a number of rules for the prevention of reddening which have been found to work well in commercial practice. They are as follow:

> 1. Floors, weighing scales, wash tanks, dressing tables and everything with which the fish comes in contact in preparation should be frequently washed with water under strong pressure.
>
> 2. The water used on fish, or in any process around the plant, should be pure. Sea water from the vicinity of any town or dock should not be used since it is always polluted.
>
> 3. The butts, or hogsheads, and dressing tables, should be cleaned inside out and steamed, or washed with a detergent.
>
> 4. Drying should be carried on as far as possible; growth is inhibited on drier fish.
>
> 5. All refuse should be removed promptly, for bits of fish act as incubators in continuing infection.
>
> 6. Fish should be so kenched in the storeroom that free circulation of air is permitted around the stack. There should be no dead air spaces.
>
> 7. All equipment should be thoroughly washed and steamed each time after use.

8. The finished product should be held in a reasonably cool place and when shipped should be handled under proper temperature conditions, as are cured meat products.

9. All buildings should have good provisions for light. Most buildings are too dark.

Fumigation with the vapors of burning sulphur or sulphurous acid, or with formaldehyde, has also been found effective in controlling the growth of red halophilic organisms, using 10 ounces (283 gm.) of formaldehyde or 5 pounds of sulphur per 1,000 cubic feet (28.3 cu. m.) of space. Fish-curing salt mixtures containing 2 percent sodium acid phosphate and 0.25 sodium benzoate have been found to stop the growth of red bacteria. Boric acid mixed with fine salt and dusted on dry salted codfish has been used commercially for about 50 years. It has some effect on inhibiting the growth of red bacteria on dry salt fish. The use of these two chemical preservative agents is permitted under the Food, Drug, and Cosmetic Act only if the amount is less than 1/10 of 1%, and if their presence is declared on the package together with directions for removal.

Dun Cod

"Dun" cod is another form of bacterial discoloration on dry salt fish. It is a type of spoilage that has also been known for a long time. This is dry salt fish on which dry, freckle-like spots have formed. These spots may be found either on the skin or flesh side and usually on fish that have been cured for some time, but may be found on comparatively fresh fish. The degree of infection varies from a scattering of chocolate brown spots to an almost complete coating of the surface (Hess 1940). The standard commercial method of control has been to scrub the fish with a brush under water, wipe off, powder the surface with a mixture of fine salt and boric acid, and dry for a day or two. Dun fish has been found to be caused by a brown halophilic mold (*Torula epizoom*). This organism has an optimum of 10 to 15 percent salt with growth over a wide pH range. Recent investigators have found that sodium proprionate is an effective inhibiting agent and is suitable for commercial use (Frank and Hess 1941). Fumigation with formaldehyde or burning sulfur has also been recommended. According to Stevenson (1899), dun fish was at one time prepared as a type of cure for markets which desired its special flavor.

Molds

Dry-salted and smoked fish are sometimes attacked by mold, particularly where the fish have been very lightly salted with most of the moisture removed by drying. Some of the molds have been identified as: *Rhizopus nigricans*, the black mold of bread, associated with the spoilage of many varieties of food; *Aspergillus glaucus,* a green mold found on hay, grain, preserves, jellies, and dried meats, and one of the most important common molds; *Mucos mucedo*, at first yellow, then becoming dark brown or black; *Botrytes grisea*, which produces a white color; and Thomnidium sp., which is characterized by an orange-yellow color. The presence of molds indicates poor storage, that is, the product has been allowed to become damp and held at too high a temperature. When these molds first appear, a commercial method of control is to rub the surface well with a vinegar-dampened cloth and then dry the fish for a few hours.

OTHER TYPES OF SPOILAGE

Dry-salted fish, if properly prepared and stored, are generally immune to the common spoilage organisms attacking fresh fish. The fish are subject to spoilage through a variety of other causes from the moment the curing process begins. These agencies may be divided into: animal life, insects, enzymic changes, and physical and chemical deterioration.

Animal Life

The most common types of destruction through animal life are by sea birds such as gulls, while the fish are being dried, and by rats or mice while the cured product is in storage. Sea gulls tear the flesh while they are on the drying racks and scatter their droppings over the surface of the drying fish. In many areas the drying racks are given special protection, such as a covering of old fish netting. Among mammals, the rat has been called "the arch enemy of the food preservationist" and different species may also be included (Rector 1925). It is said that these rodents eat half a million dollars worth of food annually in the United States, including dry salted fish, and in addition they destroy a great deal more. This topic is discussed at length in publications on food preservation, and agencies such as the U.S. Department of Agriculture and the Branch of Predator and Rodent Control of the U.S. Fish and Wildlife Service have made extensive studies with recommendations for control measures. Control can be summarized in a few words: Good warehousing and packing is probably the best single means of reducing spoilage by rats and mice.

Insects

Two types of insects in particular damage dry-salted and smoked fish. Freshly salted fish on drying racks are attacked by several species of flies, which lay eggs that soon hatch into maggots. Infestation with maggots is entirely unnecessary. It is most apt to occur where the fish are slack (light salted), not carefully dried, and where no attention is paid to the sanitary condition of the plant or drying yard. If the flakes (drying racks) are properly located and the yards kept clean, there is very little danger of maggot infestation. Flies that may appear in still, calm weather will be kept from doing damage by a smudge from green wood.

The most common of European insect parasites, present to some extent in America, is the fish mite. This is apt to appear if the drying fish become the least damp. If codfish are "mited" they look as if they had been sprinkled with coarse black pepper. Mites may attack fish either while on the flakes during the drying process, or when stacked in the storeroom when drying is completed. Mites commonly appear when a sudden fog comes up while the fish are laid out to dry and the curers cannot get them gathered up before they are damp. The drier the fish at such a time, the greater the risk. Drying the damp fish in a shed over a low clear fire for 24 hours should give sufficient protection. Some curers take the additional precaution of scrubbing fish in strong brine before they are placed outside for a final drying. To guard against infestation by mites while the dried fish are in the storeroom, the warehouse should be kept thoroughly clean, dry, and airy. The salted fish should not be stacked in the warehouse during the heat of the day. They should be gathered from the flakes either in early morning or at evening when the heat of day is past. If the climate is damp, the dried fish should be well covered in the warehouse.

Enzymic Changes

Enzymic changes are usually thought of as occurring in fresh fish. Recent studies indicate that these changes may take place in preserved fish in storage. They are not found in canned foods but enzymic deterioration has been found to take place in frozen fish stored at O° F. Enzymic changes may also take place in dry-salted fish.

Enzymes are digestive agents which break down or build up food substances. Enzymes are organic substances and to date little is known as to their exact composition. The digestive fluids of fish are very rich in enzymes. The cells in fish tissues also contain enzymes which do not cease their activity upon the death of the fish, but may digest the cells in which they are contained. This process is inhibited or greatly reduced by dehydration but it may still go on, though at a much slower rate, if other conditions, such as temperature, are favorable. For example, dehydrated fish with a good flavor, stored in a sealed metallic container, may show undesirable changes in both color and flavor after several months of storage at room temperature, or about 70° F.

About all that can be done to reduce enzymic deterioration in dry-salted or smoked fish is to dry the fish down to a low moisture content, and store them in a cool place, at an even temperature, if they are to be held for any length of time. Enzymic spoilage is comparatively easy to control in smoked, dried, and salted fish. It causes the most trouble in the preparation of dehydrated fish.

PHYSICAL AND CHEMICAL DETERIORATION

The principal physical and chemical causes of spoilage of dry-salt cod, which are also more or less important in other cured fish are: (1) oxidation; (2) light; (3) temperature changes; (4) water; (5) fire; and (6) contamination with foreign materials. This is a fairly complete outline of types of deterioration caused by physical or chemical action causing harmful changes that have been studied by chemists, engineers, and food technologists.

Oxidation

Oxidation is probably the most important chemical change which may take place in dry-salt cod and other cured fishery products. It may occur while the fish are being cured, or while they are in storage awaiting retail sale. It is discussed at some length on other sections of this report as it affects the preparation of specific products. The "sun burn" of dry salt cod, and the "rusting" of salt herring, alewives, and mackerel are instances of oxidation. Its importance may be illustrated by saying that it is sometimes necessary to discard 25 percent of mackerel that are being repacked because of oxidation changes. Sometimes an entire lot of dry-salt cod is lost on the flakes because of oxidation due to faulty drying. Oxidation occurs most rapidly and frequently in fatty fish but may occur in "lean" fish such as cod. It is characterized by the development of rancid flavors and odors, and the occurrence of undesirable color changes. Study of oxidation changes in cured fish is handicapped by the fact that these alterations are often complicated by deterioration caused by microorganisms, enzymes, and probably other factors strictly chemical in action.

Light

Sunlight is not usually considered a destructive agency; on the contrary, it is thought of as beneficial. Yet it may set in motion many chemical reactions that assist in the spoilage of cured fishery products. As a matter of fact light waves, particularly the short invisible rays called ultraviolet, are intensively active and very few organic substances can stand long exposure to intense light without some changes in composition (Rector 1925). That is the principal reason for the caution printed on labels, "Store in a cool, dark place." Light changes occur in salted cod when they are first set out to dry. Intensity of light rays is a factor. For this reason it was formerly considered impossible to obtain dried salted fish of good quality in warm climates. The effect of light is discussed in detail in the section of this report, Dry Salting Fish in Warm Climates. Light changes also occur when cured fishery products are packed in glass containers. Shredded cod packed in glass tumblers gradually turns yellow and rancid on long exposure to light. Spoilage in which light is a causative agent usually develops at such a slow rate as to be almost unnoticeable. For this reason many of the changes of which it is a factor are obscure. Light is probably not important as a direct and single cause of damage, but is so often connected with some obscure cause of spoilage that it must be carefully considered.

Temperature Changes

Temperature of storage is an important factor in all types of spoilage described so far in this discussion. The development of insect infestation, micro-organisms and enzymes, is greatly affected by temperature. True chemical changes of all kinds increase in intensity as the temperature rises. It has been calculated that the speed of reaction is doubled with each increase of 18° F. Temperature is, however, a contributory cause rather than a principal agent in these changes. There is, in addition, spoilage due directly to marked temperature changes. This may be of two types; first, changes caused by excessively low temperatures and second, those due to temperatures higher than normal.

Cured fishery products are not damaged by freezing. In fact freezing temperatures preserve dry-salt cod and other cured fishery products for much longer periods than would otherwise be possible. Some types of smoked fish, such as kippered herring, deteriorate after a comparatively short period of freezing storage, but in these instances freezing is a contributory influence, not the principal direct cause.

In most spoilage initiated at high temperature, temperature itself plays only an indirect part. In general the activities of insects, microorganisms and enzymes are intensified by temperatures higher than normal. As to direct damage, there is some evidence of loss of flavor and physical softening not due to decomposition, caused by high temperature.

Oxidation, colloidal changes, and the action of light are all speeded up by a rise in temperature. When cured fishery products are to be held for long periods, the knowledge that the speed of chemical reactions is reduced one-half for each 18° F. drop in temperature immediately suggests storage at low temperature. In the storage of hard-salted cod, it has been determined that storage at 40° to 45° F. is optimum.

Water

In methods for the preparation of dry salt cod and other cured products it has been emphasized that removal of the natural moisture of flesh is the principal requirement in preservation. The presence of natural moisture accelerates spoilage. The food technologist also has to deal with extraneous moisture which may come into contact with cured fishery products and also act as a spoilage agent.

Extraneous water usually gets into dry salt fish in one of the following ways: rain on the fish while exposed on drying racks; wetting with salt or fresh water during transportation by ship; and wetting due to defective plumbing in the storage building or as a result of attempts to put out a fire. Fire and smoke damage will be a complicating factor in spoilage in the last instance. Water damage may also be caused by a third factor, that of atmospheric humidity. If cured fish are not given good storage when built in a damp, humid climate, or if attempts are made to dry salt fish in such a climate without special precautions, the excess moisture in the atmosphere is absorbed by the product.

No matter how the water gets into the dry salt fish or other cured fishery product, the result is the same; they will mold or become slimy. In warm climates "reddening" is increased (or intensified) by high water content. If water damage is extensive the growth of spoilage bacteria is accelerated.

Fire

Fire damage may occur in any one of several forms. The cured fish may be entirely consumed or become so charred and carbonized that it is unappetizing and inedible. It may absorb such products of combustion as smoke, dust, or cinders, destroying all or part of its value. Fire protection is valuable as a spoilage preventive agent at all stages of preparation and distribution. Rector (1925) states that:

> "Animal foods of all kinds on account of their high protein content give off very disagreeable odors when burning. They are also as a class very susceptable to odors from a distance and are likely to become damaged by comparatively little exposure to smoke."

Contamination with foreign Materials

There are many opportunities for the contamination of cured fish by foreign substances. It is most apt to occur during shipment, storage, or at the retail store, for while the fish are at the plant the packer is able to control the conditions. The principal types of contamination are substances such as kerosene, lubricating oil, or other petroleum products, which may come in contact with the dry salt fish during storage or shipment; poisonous chemicals; air-floated dust; metallic contamination, especially by zinc, tin, or copper; contamination by coal and decomposing organic matter which may occur during storage or shipment. Good packaging by the packer is therefore essential. Prevention of spoilage by contamination through foreign agents may be given in two words, improved packaging. A slight additional cost will prevent the loss of entire lots of fish.

Dried salted fish may act as a contaminating agent to other food products. If the fish are exposed in the open, especially when not skillfully prepared, and have a fairly high moisture content, the strong "fishy" odor is absorbed by other foods. For this reason some grocers are unwilling to handle dry salt cod or other cured fish. The retail sale of cured fish has naturally suffered. This is an additional reason for better packaging.

Chapter Ten

Miscellaneous Dry Salt Fish

ANCHOVIES, SPANISH STYLE

The curing of anchovies centers around the Mediterranean and South Atlantic. Spain, Italy, Portugal, France, and Greece are the most important producing areas. Several species of anchovy are found in the Americas. The Pacific anchovy (*Engraulis mordax*) found from Vancouver Island to Lower California, is the largest and most valuable species in the Americas. It is a good example of our under-utilized fish species. The California anchovy is known to frequent the coast in large numbers, yet it is used commercially in only small amounts. From Puget Sound and Willapa Harbor southward, the anchovies are caught commercially at irregular intervals, but mainly during the summer months. The catch is taken largely by small purse seines, lampara, and ring nets. The fish are used mostly as bait, either alive or salted. Some are kippered or cured Spanish style, which is the usual way this species is prepared for human consumption. Anchovies salted for bait are given much the same cure as dry salt herring.

The preparation of Spanish style anchovies is carried on mostly at Monterey, although occasionally small lots are packed at San Pedro.

The freshly caught fish are first gutted and headed. The workmen do this by pulling the heads off with their fingers. The anchovies are dropped into three-quarter or half-ground salt in which they are thoroughly mixed, then packed in salting tanks or large butts. Accurate data are lacking as to the amount of salt, but it is reported that it is from 25 to 30 pounds per 100 pounds of fish.

In order to obtain the proper cure, the fish must be exposed to warm temperatures. Therefore, the tanks are held at temperatures varying between 80° and 90° F. for a period of about 4 months. They are ready for use when the flesh has a red color from skin to backbone, and a strong, sharp flavor. The anchovies are removed from the tanks and drained for several hours, then packed in round tins holding 8, 14, or 28 pounds. The fish are pressed to remove excess moisture and oil. When the proper weight has been filled in, the cans are sealed, and are then ready for shipment. The cans are not processed or pasteurized, and preservative agents are not used.

Barracuda

The Pacific barracuda (*Sphyraena argentea*) is dry salted commercially only in southern California. At one time is was a regular item of some importance, but now it is cured only when there is a surplus of barracuda on the fresh-fish market. In some years no production is recorded. There is usually some small noncommercial production by the fishermen for their own use. There is also some dry salting of barracuda in Puerto Rico and Florida for home consumption. When properly cured, the barracuda makes a very good dry salt fish, white and dry, and of excellent flavor; however, some of the product is dark in color and has a strong flavor, probably due to faulty methods of curing.

The recommended method for curing barracuda is as follows: The head is removed, leaving the collar bones. A slit is made down the middle of the belly to the vent, and the body is cleaned thoroughly. A cut is made just above the backbone on the abdominal side, cutting along a line where the ribs join the backbone, and continuing almost to the tail. A similar cut is made just below the backbone. A sweep of the knife through the thin section of rib bones still adhering to the flesh removes the backbone, which is broken off near the tail. These cuts must not reach through to the skin. When split the barracuda should lie flat in a single piece. After splitting, the fish are washed thoroughly in salt brine and soaked 30 minutes to remove all traces of blood. The flesh is scored almost to the skin, the cuts running longitudinally from collar bone to tail at approximately two-inch intervals.

The fish are allowed to drain, then roused (dredged) in a shallow box of packers fine or three-quarter ground salt. The fish are picked up with as much salt as will cling to them and packed flesh side up in tanks or tubs. Sufficient salt to cover any exposed surface is scattered over each layer of fish. As a general rule, approximately one part of salt should be used to four parts of fish. A loosely fitting cover is placed on top, with sufficient weights to keep all fish under the surface after the brine has formed.

After 48 hours in salt the barracuda are removed, scrubbed well in salt water or brine to remove any slime or excess salt, piled into heaps to drain for 20 or 30 minutes; then laid on racks to dry. The racks are usually wooden frames with legs 3 feet high and a top of reed, bamboo, or galvanized chicken wire. If drying is carried on under the direct rays of the sun, oxidation or rusting sets in immediately. However, if the barracuda are kept shaded, in a breezy location, they will dry well, with a clear color, and will not oxidize readily afterward. The fish must be turned frequently during the first day of drying. At the end of the day the fish are stored under cover in a dry shed. They are weighted to press out additional moisture. The next morning the fish are again laid out to dry. After five days of good drying weather the barracuda are usually sufficiently dry. If poor drying weather sets in, the barracuda are restacked in the storage shed and weighted very heavily.

Herring

Before World War II, the largest production of dry-salted herring was in British Columbia, where large amounts were prepared for export to the Orient, especially China. Production in British Columbia has decreased to a small fraction of the former level. It is not known whether postwar conditions will call for large-scale manufacture of this article. Dry salt herring has never been of more than minor commercial importance in the United States. The production of dry salt herring in this country has varied between 150,000 and 200,000 pounds annually.

As packed on the Pacific coast, dry-salted herring is a very roughly prepared product. The Pacific herring (*Clupea pallasii*) are not gutted, scaled, or washed. As the fish are landed, they are shoveled into a heap of salt, turning the pile over with a large scoop shovel until herring and salt are thoroughly blended. The mixture is then packed in large tanks with a capacity of 100 barrels or more. Additional salt may be added to the tanks. The herring are allowed to cure for several days, forming their own brine. They should be left in salt for 10 days, but if time is short, they can be removed in from 3 to 7 days. They may be left in salt up to two or three weeks.

When the packer is ready to make a shipment, the herring are shoveled out of the tanks onto the floor in heaps which are covered with the excess salt from the tanks. The fish are allowed to drain for 1 or 2 days. Then they are shoveled into large wooden boxes similar to those used in shipping fresh halibut or salmon. Some salt is added between the shovelfuls, the amount depending on the length of time the herring have been in the salting beds. A scattering of salt is spread over the top of the fish, the cover is nailed on the box, and it is ready for shipment. Dry salt herring is reported to be valued in the Orient largely for its salt content, being used for seasoning of food such as stews.

Mackerel

In Spain, southern France, and on the coast of North Africa, especially Algeria small mackerel are cured according to the same method used for anchovies. Fish of a size fro 15 to 20 to the kilogram (2.2 pounds) are used; that is, mackerel averaging about 8 to the pound. On arrival at the plant, and before heading, the mackerel are placed in tanks without brine, the bottom of the tanks sprinkled with salt. The baskets or boxes of fish are thrown into the tank by hand to form thin regular layers. Handfuls of salt are thrown over and among the layers of fish, so that each layer has a thin covering of salt.

Brine begins to form through extraction of moisture from the fish by the salt. In about half an hour a saturated brine is formed in which the fish are almost floating. The fish are washed in this brine carefully, one at a time. As they are washed they are placed in small baskets for drainage, and then spread on tables for heading and evisceration.

When the mackerel are dressed they are replaced in the baskets which hold 8 to 10 kilograms (17.6 to 22.0 lb.) each. Filled baskets are emptied into a washing tank which is one-third full of fresh 25° B. (100° salinometer) brine. Some salt is scattered through the fish as they are thrown in.

As in preparing anchovies, it is desirable to have several tubs full of fish before beginning packing so that this operation will not have to be interrupted. The first step is to wash the mackerel thoroughly, after which they are placed in baskets to drain before being placed in the barrel.

Enough "demi-fine" No. 2 salt (three-quarter ground) is scattered on the bottom of the barrel to cover the wood. The tiers are not in rosettes or transverse layers as is the custom of packing anchovies and sardines, but each is a series of straight parallel rows, with each row containing one, two, or three fish according to its place in the layer.

A single fish is laid on its side, the back against the wall of the barrel. Two fish are then packed in with backs against the belly of the first fish, heads against the sides of the barrel, and tails crossed. This is continued until the last row in the layer is reached. The single fish in this row is packed belly to belly against the preceding fish in the next row with its back against the staves of the barrel. The gaps left where the tails of the fish are crossed are filled by fish laid alternately with the heads to opposite sides. The tails are hidden under preceding fish in such a fashion as not to be apparent. A very thin layer of salt is then thrown over the fish, and the packer is ready for the next layer. It is important to use careful packing, keeping the layers even and the fish straight in the rows.

When the barrel is filled to the rim, four or five additional layers are packed above the top. The last layer is given a slightly heavier coating of salt, the lid is placed on top and loaded down with weights to an amount of 40 or 50 kilograms 88-110 lb.). This is called the "first filling." The next day the brine which has formed is drained off and discarded, and the top layer is washed until the fish are bright and clean. The barrel is then filled with mackerel of the same lot. This is the second filling. Three days after the second filling the process is repeated for the third filling.

Two days after the third filling the brine which has collected is again drained off, the top layer is washed and covered with a layer of quarter-ground salt to a uniform depth of 1/2 centimeter (3/8 inch), the cover is placed on top, and the barrel is set on a rack in the fish house to cure. Some fresh brine is poured into each barrel to make up for the brine which was drained off. The barrels must be inspected for leaks from time to time while they are undergoing the curing process.

The ripening is finished in from 2 to 3 months. It is much more rapid than that of the anchovy and the sardine by reason of the much greater quantity of blood contained in the mackerel. When the fish are well matured, the flesh is a beautiful red, similar to that of the bonito, and liberates a characteristic agreeable anchovy odor. They are eaten as hors d'oeuvre (Dieuzeide and Novella 1942).

Mullet

The striped mullet (*Mugil cephalus*) is the most important dry salted fish in the South Atlantic and Gulf of Mexico areas. The principal area of production is from Carabelle to Cedar Key on the gulf coast of

Florida, with most of the production at these two ports, and in a few other scattered localities.

Most of the mullet are taken by run-around gill nets with a smaller quantity caught by haul seines. A great deal of the mullet must be transported a considerable distance to the curing plant. Some are still brought in by boat but probably the greater share is now carried by truck. No special care is used in handling. As a rule the fish are dumped in the body of a truck, without ice. When delivered, they are frequently mixed with sand and weed. The fish should be washed before loading in the truck or boat and given better care in handling.

Preparation begins as soon as the mullet are unloaded at the fish house. They are first washed to remove sand and other debris. The heads are usually removed, especially from the larger fish, but the collar bones or napes are retained, since if this is not done, the fish are apt to shred in subsequent handling. The mullet are split in a manner usually referred to as mackerel style. The splitter holds the fish in place on the splitting board with his left hand, with its tail toward his body. He places his knife blade so that it will scrape along the backbone, then splits the fish along the back from head to tail without cutting through the belly, leaving the backbone on the right side. The cut is not completed, and two halves are left attached at the tail of the fish. The backbone may be partially removed in the larger mullet by making an additional cut on the underside of the vent, thus cutting the backbone loose so that it can be broken off for about three-quarters of its length.

If the mullet contain roe, the sacs are carefully removed for salting separately. The viscera, coagulated blood, and membrane, or so-called black skin lining the body cavity are scraped out as thoroughly as possible. The more careful salters scrub the intestinal cavity with a piece of gunny sacking or coarse canvas, the most effective means of removing black skin and blood. The flesh is scored longitudinally on both sides at intervals of about one inch, so that the salt may readily penetrate the flesh (Jarvis 1945c).

The fish are then thrown into a tub of clean sea water, where they remain for about 30 minutes, after which they are thoroughly washed. This soaking also aids in removing diffused blood from the flesh. After the mullet are removed from the washing tank they are allowed to drain for a short time to remove surplus moisture. The fish are then rubbed thoroughly with a medium-grain salt, usually a Louisiana mined salt. They are next piled in kenches or stacks inside the fish house. The kenches are about 3 feet high. The fish are arranged in regular layers, skin side down and two deep, with heads toward the outside and tails to the center. A thing layer of salt is scattered over each layer of fish. The total amount of salt used is about 30 pounds to 100 pounds of fish.

The mullet remain in the kench until they are to be placed on the market, which may not be for three or four months. Some packers weight down the kench to press out more moisture, and may re-kench the fish once or twice, reversing the order of layers, but this practice is not common. For shipment, the fish are brushed, and weighed, then placed in open boxes or bushel baskets with a scattering of salt on top.

The following method is used by those who wish to prepare a really choice product and are willing to go to the extra trouble. The cleaned fish are washed thoroughly, and dropped into a tub of salt brine made in the proportion of one pound of salt to one gallon of water. They are allowed to soak in the brine for 30 minutes to remove all traces of blood from the cut flesh. After brining, the mullet are drained for at least 20 minutes to remove surplus moisture.

A shallow box, about 2 feet square, is filled with salt, usually a dairy-fine grade. The drained fish are rolled in salt, which is also rubbed into slashes in the flesh. A thin layer of salt is scattered over the bottom of a tub or salting tank. The mullet are then picked up with as much salt as will cling to the body and are packed in even layers in the tub, flesh side up with each layer at right angles to the preceding one. A small amount of salt is scattered between layers. A loose-fitting cover is placed on top which is weighted down enough to cover the mullet by brine formed. In warm weather, a saturated brine may be added immediately, instead of allowing the brine to form gradually by extracting moisture from the flesh. he amount of salt used averages 25 pounds to 100 pounds of fish. This is about 17 percent less than that given previously for kench curing mullet.

The mullet should be sufficiently salted in about 48 hours, after which they are removed from the brine. The fish are then scrubbed thoroughly to remove any traces of excess salt and placed in layers, flesh side up (except for the top layer) on a low rack. The stack is weighted down heavily to press moisture out of the fish. The next morning the mullet may be hung in a shady spot where there is a good breeze, or dried on racks of chicken wire in an open-walled shed. At night they are re-stacked and weighted down, but are set out to dry again next morning.

In good weather the mullet will be sufficiently cured after 6 days of drying. In unfavorable weather, and for larger fish, more time is required. In damp, humid weather the salters allow the fish to remain kenched, restacking them occasionally with a scattering of salt between the layers. After the drying process is complete, each fish is sprinkled with fine, dry table salt, wrapped in waxed paper, and packed in lightly covered wooden boxes holding about 10 pounds.

Kingfish (King Mackerel)

Small quantities of king fish, also called king mackerel (*Scomberomorus cavalla*) are dry-salted in the Florida keys and on the west coast, for local sale. The quantity so prepared has decreased in recent years, but this is still a commercial product.

The fish to be salted are usually those taken during the first part of a long fishing trip, or those which cannot be disposed of profitably in the fresh-fish market. The fish are split open and eviscerated. As a rule the backbone is cut out. The membranes and blood are removed and the fish washed indifferently. The thick flesh is slashed crosswise from back to belly, with cuts about 2 inches apart. A small amount of salt is sprinkled over the surface of the flesh and rubbed well into the cuts. If the fish are to be used in a day or two, they are given no air drying, but if held for a longer period they must be dried. The fish are spread on shed roofs or woodpiles, flesh side up, or are hung under the eaves by their tails. If they are not hung, the kingfish must be turned frequently during the drying period. The amount of drying given depends on the individual curer.

Red Drum (Channel Bass)

The red drum or channel bass (*Sciaenops ocellata*), is caught in considerable numbers along the South Atlantic coast in the late summer and fall. The greater part of the catch is made with runaround gill nets and handlines, in almost equal amounts. A lesser quantity is taken by haul seine and a few by trolling. As there is not a great demand for this fish in the fresh-fish market, some of the catch is dry-salted for local use, especially along the North Carolina banks. The only market is among the coast people, the fishermen, and their neighbors on the shore of the adjacent mainland.

The heads are chopped with a hatchet, the fish are split into two sides, and the backbones are removed. Each side is scored through the flesh lengthwise from collar to tail, the cuts penetrating almost to the skin, and about two inches apart. The sides are washed thoroughly in sea water or brine, to remove all traces of blood or other waste, and are drained for about 20 minutes.

The drained sides are rolled in a box of fine salt, which is rubbed well into the flesh, especially into the cuts. The sides are then packed in tubs, flesh side up, in even layers. A little more salt is scattered over each layer and the top is weighted down. The tub is then filled with saturated brine. The fish are allowed to remain in brine for about two weeks. The sides are then removed and scrubbed thoroughly to rid them of any excess salt, blood spots, or black skin. The sides are stacked not more than 2 feet high, the bottom layer being placed skin side down, and the others skin side up. The top is covered by boards and weighted down with rocks.

The next day the sides are hung in a shady place for about 10 hours of air drying. If the weather is unfavorable, the kench is re-piled, reversing the order of layers. The flesh must not be exposed to direct sunlight during the first few days of drying, or a crust will form which hinders the removal of moisture from the inner flesh. Oxidation is also more likely to occur. At the end of the day the fish are stacked as before and weighted down more heavily. They are left in the kench during the next day, then given a day of drying, followed by a day of pressing. This is continued, increasing the weights each time the fish is pressed, until the fish are well dried. The product is considered thoroughly cured when the pressure of a thumb in the thick part of the back leaves no impression.

Salachini (Salachi)

This fishery product is specially favored by the Italian and Greek trade. Large amounts are prepared in the Mediterranean area where the product is believed to have originated. It is also packed in California in the Monterey Bay area. Occasional packs are sometimes prepared in the Los Angeles harbor area. The English name is dried salted pressed pilchards.

Only the pilchard (*Sardinops caerulea*) may be used in preparing salachini. There is no selection of raw material, in that all sizes are used, but the fish must be fresh and firm.

The pilchards are salted in the round in tanks, using California half-ground fishery salt, which is mixed with the fish. Additional salt is scattered on top and the brine is allowed to form from the fish. The proportions of fish and salt have never been determined accurately, but it is estimated that 35 pounds of salt is used to 100 pounds of fish. The pilchards are left in salt for at least three weeks, sometimes longer. When they are to be prepared for shipment they are removed from the tanks and allowed to drain. If there is any excess salt on the fish it is removed by washing in brine.

When the pilchards are sufficiently drained they are packed in small tubs, 18 inches in diameter by 12 inches high. They are not gutted, gibbed, or headed. The fish are packed in tubs rosette style, with heads to the outside and tails to the center. About 6 additional fish are required to fill the center. Approximately one thousand fish, 7 to 8 inches long, are packed in a tub. A collar of tin is fixed around the top so that the fish can be packed above the level of the tub top. When fish have been packed in to the level of the collar top, a cover slightly smaller than the tub top is set in place, and the fish are subjected to heavy pressure. This is usually done by setting the tub under an overhead beam, then fixing a jackscrew in place, and turning the screw. After the sardines have been pressed down, the tub must be refilled once or twice. Two hundred pounds of salted fish are reduced to approximately 75 pounds of finished product by this method. The tub is not yet tightly coopered so that the brine and a good deal of oil may run out of the bottom and between the staves. The oil is collected and sold.

Salachini is also packed in boxes and tubs of other sizes. It is characterized by the more or less red-color flesh and rather sharp, pungent flavor.

The method given for preparing Salachini is similar to that used for many years in Cornwall for preparing pilchards for the Italian market under the name "fumadoes." It is described by Holdsworth (1877) as follows:

> "The curing is the especial work of the women, who pack the pilchards in alternating layers of coarse salt and fish on the stone floors of the curing house until the "bulk" has reached a height of 5 or 6 feet. Here the fish remain for a month, and the oil and brine draining from them are carried off by gutters in the floor to a cistern. When the fish have been sufficiently salted, they are washed and packed in hogsheads, each layer of fish being packed with their heads outward and with a "rose" of fish in the center. A circular piece of wood called a buckler and rather smaller than the head of the cask, is then placed on the top of the fish and a strong but gradual pressure is applied by means of a lever until the mass of fish is reduced one-third in bulk and a great quantity of oil squeezed from them.
>
> This drains through the sides and bottom of the cask, the hoops of which are not at that time tightly driven, and is collected as before. The quantity of oil obtained from the pilchards depends on the season, but at least two gallons of oil are expected from each hogshead.... The cask is filled up three times before the pressing is finished, which

is not until after 8 or 9 days, and then the hogshead (50 gallons) of fish should weigh 4 cwt. gross. The average number of fish packed in a hogshead is about 2,500.

As the name of this product indicates, the fish were originally smoked. Now salting and pressing alone are used for curing purposes. The salted fish are said to keep longer and have a better flavor than the smoked pilchards.

Salmon

Before World War II large quantities of salmon were dry-salted every year on the Pacific coast, mostly for export to the Orient. With the restoration of normal trade conditions, it is hoped that this trade can be resumed. The greater proportion of the pack has been prepared in British Columbia. Little interest in this method of curing salmon has been shown in the United States for a number of years, but occasionally when chum salmon are in demand for canning purposes, considerable quantities are available for dry salting at a price which should show a profit. Statistics for 1940, the last prewar year when conditions approached normal, give the United States production of dry-salted salmon as 29,000 pounds.

Chum (dog) salmon (*Oncorhynchus keta*) is used largely in the preparation of dry-salted salmon though other species are sometimes used. In Siberia, where an increasing amount of dry-salted salmon was put up every year before the outbreak of World War II, red (*O. nerka*) and coho (*O. kisutch*) salmon were used to some extent. Many of the fish are taken by gill net, though salmon caught by other gear may be used, especially when the market is glutted.

In preparing dry-salted salmon, the fish are headed leaving the collar, split down the belly, and eviscerated. The blood is scraped out as thoroughly as possible and if large, the fish are split into sides. he backbone may or may not be removed, depending on the custom of the original curer. As a rule, it is removed when the salmon are split into sides. Small fish may be split almost through to the skin, but are left in one piece and the backbone is not removed. The dressing and splitting process is the same as that described for hard-salt or pickled salmon but is carried out more roughly after splitting. The fish are washed and given a final cleaning in the washing tank to remove bits of viscera or other offal overlooked previously.

When the salmon are well drained after washing, they are laid down in stacks with a layer of salt between each layer of fish. The salt must be thick enough to cover the surface of the salmon completely. All layers are piled flesh side up, with the exception of the top one, which is placed skin side up for the purpose of better protecting the fish against dirt or other contamination. Sometimes the fish are cured in vats or tanks as in pickling salmon. The brine is formed naturally and the amount of salt required is somewhat smaller than normally used in curing fish. If salted in tubs or vats the salmon are arranged as neatly as possible, flesh side up, alternating heads and tails, with the thick edge of the side toward the side of the tub, small pieces being packed in the center to make the layer even. If cured in stacks or kenches the salmon are laid down in rows, alternating heads and tails. The amount of salt required in dry-salting salmon is approximately 35 to 40 pounds per 100 pounds of fish, for the kench cure; 25 to 30 pounds for the tank cure (Jarvis 1936b).

When the salt has struck through, that is, when the fish appear thoroughly cured and it is determined that the salt has penetrated completely, they are ready for packing. An average of two weeks in salt is required. If the curer is not ready to make a shipment the fish may be left in salt longer.

Dry-salted salmon are packed in large wooden boxes holding 400 to 500 pounds, with salt scattered between each layer of fish. No particular system is followed in packing except that the packers try to make the layers even, without large air spaces. From 10 to 15 pounds of salt per 100 pounds of fish will be used in repacking. the product receives no further processing but if it is to be held any length of time before shipping, the boxes should be again repacked. Dry-salt salmon is reported to be valued for its salt content in China where there is a high tax on salt. It is sometimes added in small pieces to soups and stews for this reason, but a great quantity is eaten raw as a relish. Dry-salted salmon have been a favorite New Year's gift in Japan.

Dry Salted Atlantic Salmon

Some Atlantic salmon (*Salmo salar*) is dry-salted in Canada and Newfoundland by a method resembling the Scotch and Norwegian cures for kippering salmon as described by Duthie (1911). The method was probably introduced from Scotland.

The fresh salmon are chilled in ice until the flesh is firm. The fish are headed leaving the collarbone. Smaller salmon are split down the back to lay flat in one piece, and from three- to four-fifths of the backbone is removed. Large fish are split into sides, removing the entire backbone. The viscera, all traces of blood, membranes, or other offal are removed and the fish are washed thoroughly, then left to drain.

A curing mixture is made up in the following proportions which is sufficient for about 10 pounds of fish: 2 pounds of packers fine salt; 1 pound of brown sugar; 1 ounce of saltpeter; 1/2 ounce whole black peppers, 1/2 ounce whole cloves; and 1/4 ounce bay leaves. The spices are crushed and mixed well with the other ingredients. The mixture is rubbed into the flesh of the salmon, placed in a tub, flesh side up, with more of the curing agent scattered over each layer.

The salmon are left to cure for two to three days, then removed, rinsed, and wiped clean. Three or four thin double-pointed sticks are fixed through the skin at intervals along the back of the salmon spread out flat. They are then hung in a shady place with a good breeze until dry. This will require about a week to 10 days. The fish are not brought in at night unless the weather is damp.

This salmon is sliced thin for use in sandwiches or as an appetizer. Stevenson (1899) reports that brandy is sometimes poured over the dried fish and allowed to soak in to improve the flavor.

Shark

The dry-salting of shark is carried on commercially in Florida, Cuba, Mexico, the Bahama Islands and some of the other West Indies and on both coasts of South America. Shark is also dry-salted in the Orient and in the South Pacific. Methods differ in some details in each locality where this fish is cured, varying according to personal ideas of the curers and differences in local conditions. The method described here is representative of curing practices in the United States, but may not be the only one in use.

Curing must begin within the shortest possible time after catching, as spoilage occurs more rapidly with the shark than with many other species. The shark is first gutted and skinned, after which the carcass is split into two sides, removing the backbone. The large streak of dark meat along the midline of each side is then cut away, dividing each side into two fillets of light-colored flesh. These fillets may be further divided into two or more pieces each, if the shark is very large. A single fillet or piece should not weigh more than 3 or 4 pounds. Each piece is then scored on both sides, lengthwise, with a knife. The pieces are dropped into a tank of clean sea water or saturated brine, to soak for about an hour (Jarvis 1943a).

The fillets are drained of excess moisture after brining, which requires 30 minutes. They are then rolled around in a shallow box half filled with packers fine salt, which is rubbed well into the slashes in the flesh. The fillets are packed in a tub, in layers, with each layer at right angles to the preceding one. Salt is scattered between each layer of fish. The top is weighted down to keep the flesh below the surface of the brine which forms. The shark fillets remain in salt from 2 to 7 days, depending on the size and weather. During unfavorable weather they may be kept in salt for 10 days or possibly even longer.

When the meat has been sufficiently salted, the pieces are scrubbed thoroughly in fresh brine, and laid in small heaps to drain for 2 or 3 hours. They are hung out to dry in a shady location having a good breeze, or are laid out on drying racks covered by a roof as a protection against the sun. Drying under the direct rays of the sun is apt to discolor the flesh, especially during the first days of the drying period.

At the end of each day the shark fillets are stored in a dry shed or warehouse, piled up in small heaps with weight on top equivalent to about half the weight of the fish. The next day the fillets are again dried and in the evening stacked under weights with the pressure increased somewhat. The alternate drying and pressing are continued, with the pressure constantly increased until the cure is complete. This requires about 10 days (Caribbean Commission 1945).

If drying conditions are not good at the end of the drying period, the fish may be hung in a light smoke for one day (about 10 hours). The temperature should not be much over 80° F., and the density should not be sufficient to color the product to any extent. The fillets are packed in shallow wooden boxes, tightly covered. Sometimes each fillet is wrapped in paper or the box may be paper lined, but this is not the general rule.

Tuna

Dry-salted tuna has always been an important Japanese fishery product. It is prepared in large quantities in Japan, to some extent in the South Pacific, and in California, being introduced into both of the latter areas by the Japanese. Either bluefin (*Thunnus thynnus*) or yellowfin tune (*Neothunnus macropterus*) is used, but the yellowfin is preferred. This method is generally utilized in California to preserve catches from the fresh-fish market.

The tuna are eviscerated and headed, then washed thoroughly. They are then cut into sides, removing the backbone. Each side is again divided into two tenderloins or fillets, cutting away the dark flesh along the lateral line. The fillets are washed thoroughly in clean sea water and are then soaked in brine for about two hours. The purpose of brining is principally to remove diffused blood from the flesh. The brine is believed to test about 50° salinometer. After brining, the fish are drained and each fillet scored with a number of longitudinal cuts on each side. If they are large, the fillets may be subdivided into two or more pieces each, cutting both longitudinally and in cross sections. If a fillet is too thick it will not cure properly.

The individual pieces are rolled in salt, rubbing it well into the cuts or slashes in the flesh. Any salt readily available is used. The pieces are picked up with as much salt as will cling to the flesh and packed in layers in tubs or salting vats, with an additional scattering of salt over each layer. They are left in salt from 5 to 10 days, depending on the size of the individual pieces, and on weather conditions. The fillets may be kept in salt longer than 10 days if there is an extended period of unfavorable drying weather.

When the pieces are considered struck through, that is, when the salt has penetrated the flesh completely, they are washed thoroughly in clean sea water and laid out to dry in a shady location exposed to a good breeze. If dried under the direct rays of the sun the color will be darker, the flavor stronger, and the product will not keep so long.

The fish are gathered up every evening so that they will not be exposed to the night dampness. Some curers stack the fish in small heaps at night, and these are weighted down. This presses a small additional amount of moisture from the fish. From one to two months are required for drying. When completely dried, the tuna are sprinkled with fine salt, the pieces are wrapped in paper and packed in light wooden boxes.

Mojama

In Spain, salted and dried tuna is prepared under the name of "mojama" (Uriarte 1926). The lean meat from the area of the backbone of unspawned tuna, called "descargamento," or any portions of the flesh of spawned tuna except the belly flesh, is cut into strips 5 centimeters (2.0 in.) thick and 50 centimeters (19.7 in.) long. The dark meat along the lateral line is trimmed away. The fillets are washed more or less thoroughly in sea water. After draining off the surplus moisture they are rolled in coarse fish salt (usually Cadiz), then salted down in kenches about 1 meter (3 1/4 ft.) high, with salt scattered over each layer of fish until covered. The tune are left in salt from 24 to 36 hours. They are then washed thoroughly in clean sea water, and any salt crystals are scrubbed from the flesh. After washing four times, the strips are allowed to soak for a few minutes and then drained.

A sharp stick (usually a sliver of split cane) is run through the flesh of each strip near one end. A string is tied to the stick on each side and the strips of tuna, called "mojama," are hung from the loop formed, from overhead poles running horizontally across the open air drying ground. A temporary roof of rush mats is usually erected over the drying rack. The mojama must be dried in the shade, as it turns rancid almost immediately if dried under the direct rays of the sun. The mojama is allowed to hang until it is perfectly dry, which may take two or more weeks. Dieuzeide and Novella (1942) give a minimum drying time of one week. When the pieces are sufficiently dry they are taken down, trimmed of any ragged edges, and packed for shipment in paper-lined wooden boxes.

Mojama of good quality will have a black-red color and a fresh small, not the pungent odor characteristic of the anchovy cure. It is considered a great delicacy as an hors d'oeuvre. Mojama is served cut in thin slices together with sliced Spanish onions. A little olive oil may be poured over the mixture.

Chapter Eleven

DRY SALTING FISH IN WARM CLIMATES

PREVIOUS STUDIES

In subtropical or tropical climates it is difficult to cure fish, especially if it is to be kept for any length of time. Yet it is in these areas that there is the greatest need for protein, and fish is the best potential source. The natives of the Asiatic coasts, especially China and Japan, have made some progress in developing fish curing under subtropical conditions, but have not solved the problem. The solution of the problem of fish curing in warm climates is considered of great economic importance, for it would raise the nutritional standard of the people and also increase employment.

The Bureau of Fisheries (now the Fish and Wildlife Service) made extensive researches on the problem of fish curing in warm climates. The first by Tressler (1920), paid special attention to the physical and chemical factors involved; later Jarvis (1932a) studied operational problems in the field. As a result of these studies, it was found that it was possible to cure fish of good quality that will keep for an extended period in tropical and subtropical climates. But, it is not always practicable to create a fish-curing industry of considerable size under tropical conditions. A small local surplus may be preserved, however. In some localities, especially on the coast of Brazil, Chile, Venezuela, and possibly in sections of Africa there may be some prospects for fish-curing on a commercial scale.

Location Requirements

The method developed for fish curing in warm climates is a variation of dry salting. It is successful only when the essential points are followed closely, and if great care is given to every step in the manufacturing process. The location for the drying racks should be on dry ground on a point where there is a good sweep of breeze. All vegetation should be cleared away and the ground covered with gravel. Large areas of nearby swampy land are undesirable, but small swampy patches may be filled in. here should be an ample supply of fresh water of good quality. The use of salt water dipped up along the shore is undesirable since it is almost always heavily contaminated. Sometimes, however, it is necessary to use sea water, then all possible sanitary precautions should be taken. There should also be adequate measures for the prompt disposal of all waste, which should be burned, if possible, or at least be buried immediately.

Species Suitable for Curing

Not all varieties of fish are equally suitable for curing. In general, the white-fleshed, nonfatty fish give the best results. In experimental work conducted by the Fish and Wildlife Service the best products were made from barracuda, parrotfish, snappers, and shark. Bonito and crevalle, or carang, make dry- salted products possessing good texture, flavor, and odor. The only objection is that the flesh is rather dark colored. Fish such as the corbina and robalo make a good product, but are usually in such demand in the fresh fish market that they are not always available for curing.

Handling Raw Material

Careful handling is required from the moment the fish are caught. If possible they should be packed immediately in ice in the round (ungutted), using about 50 pounds of finely crushed ice to 100 pounds of fish. Since ice is often unobtainable in fishing villages in warm climates, the fish may be stored in a box with a top, and possibly two sides of heavy burlap, or similar material. The burlap is kept moist by dripping water on it occasionally until the catch can be landed. The evaporation produces some degree of coolness.

Gutting is advisable only if done thoroughly, scraping out all traces of blood and viscera, rubbing the belly cavity with a small amount of salt, and scattering salt over the skin. Gutting, in which the false kidney and some remnants of viscera are allowed to remain, does not reduce the rate of spoilage.

The fish should not be allowed to remain under the direct rays of the sun without protection. If tarpaulins or other covers are used, they must not rest directly on the fish. Air circulation is essential. It is also necessary to begin curing within a few hours after the fish are caught, especially in the daytime, and where ice is not used. In some instances, spoilage has been noted by the author within 3 hours after catching.

Dressing and Splitting

The fish should be dressed as landed. First they are washed in a 40° salinometer brine. If a salinometer is not available a solution is made of one part salt to 9 parts water is roughly equivalent. Washing is necessary to remove slime, loose scales, sand, seaweed, or other foreign material. The fish are then headed, gutted, and split after the method used for cod, but first the dorsal fin is removed from most of the larger fishes. This is best done by making a long cut through the flesh, along the entire length of the fin, and just at its base, on both sides. One end of the fin is then grasped and the fin is pulled away as a whole, with the short bones attached to its base. This is much quicker and easier than sawing the base of the fin. In most cases the other fins should also be removed. In cutting off the head the line of the gill flaps is followed. The napes or collar bones are not cut off, as this increases loss of weight, makes the fish harder to handle, and detracts from its appearance. A cut is then made just above the backbone, on the abdominal side, cutting along a line where the rib bones join the backbone. This is continued almost to the tail, holding the edge of the knife blade at a downward slant so that no flesh will be left on the backbone. A similar cut is made just below the backbone. A Sweep of the knife through the rib bones adhering to the flesh removes the backbone, which is broken off near the tail. These cuts must not reach through to the skin. When splitting is completed the fish should be in a single flat piece. Some of the largest fish, as for example the robalo and barracuda, may be cut into two separate sides. After dressing and splitting, all traces of blood or membranes are removed carefully. The flesh is then scored to a depth of about one-half inch, the cuts running longitudinally from napes to tail at distances of from 1 to 2 inches apart.

Smaller fish, such as mullet and Spanish mackerel, are split mackerel style; that is, the ungutted fish are split down the back just above the backbone. Sometimes the cut is made through the head, which is allowed to remain, but is cleared of gills or other offal. Sometimes, especially for larger mullet and mackerel, the head is first removed. When the knife is drawn toward the tail, it must not go clear through the skin on the other side, so that the fish is cut in two pieces near the tail. The fish should lie flat in one piece. In splitting Spanish mackerel and other fat fish, the backbone is cut nearly to the tail, where it is broken off. The "black skin" found in the bellies of mullet and some other fish is best removed by scrubbing with a piece of coarse canvas or sacking.

Brining and Salting

The fish are soaked for 30 minutes in a 40° salinometer brine, then scrubbed and drained. The principal object of brining is to remove traces of diffused blood from the flesh but it also makes the flesh a little firmer and easier to handle. A third reason is that the fishes of warm climates have a quantity of slime, which is "cut" better by brine than by fresh water. At this point the fish are inspected for proper cleaning. If the backbone has been allowed to remain, a slash is now made just under the bone to obtain better penetration of the salt.

When the fish have drained from 15 to 20 minutes after soaking, they are ready for salting. The fish are dropped singly into a shallow box about 3 feet square and with sides 6 inches high, the dimensions varying according to the average size of the fish. Salt is rubbed into the flesh and the fish are dredged in the salt, paying particular attention to filling all cuts or slashes with salt. They are then lifted up with as much salt as will cling to the bodies, and packed in a vat or tub, skin side up, arranged in even layers with a scattering of salt on each layer of fish. Each layer is packed at right angles to the one preceding. The top layer is packed skin side up and more salt is scattered over this layer than the others. The total amount of salt used should be from about 30 to 35 pounds a 100 pounds of fish. Fat fish require somewhat more salt than lean; large, thick fish more than small. In salting the fish a dairy-fine, or packers fine mined salt, with the lowest possible percentage of impurities, has been found best. Common sea salt is more apt to cause reddening, but if it must be used, it should be heated thoroughly by baking. Coarse salt pits the surface of the flesh, and does not dissolve so rapidly.

Brine forms naturally through the extraction of moisture from the flesh. Weights are placed on top to keep the fish under the surface of the brine. A saturated brine should be added if sufficient brine to cover the fish has not been formed in 3 hours.

Drying

The salting shed should be light, dry, and cool as possible. The fish are kept here in the salt brine for a period of 48 hours. The fish are not "struck through"; that is, the salt has not penetrated completely, but after considerable experimenting with varying periods it was found that fish absorb sufficient salt for curing purposes in an average of two days. Some fish, especially the smaller ones, such as mullet, require only from 24 to 36 hours in salt. On removal from the salting vats, the fish must be scrubbed in brine to remove all traces of excess salt and dirt. After draining for 20 or 30 minutes the fish are ready for the drying racks. These are frames of wood, covered with chicken wire, if available; if not, with an openwork screen of bamboo or cane.

Oxidation, rusting, or sunburning sets in immediately in warm climates if drying is carried on under the direct rays of the sun. However, if fish are kept shaded, in a breezy location, they will dry well, with a clear color, and will not oxidize readily afterward. For this reason a roof must be erected over the drying racks and the whole drying area located so that the maximum amount of breeze will sweep across the fish (fig. 17). The fish are first laid out skin side down, but are turned at hourly intervals or at least three or four times during the first day of drying.

Handling During Drying Period

The fish are gathered up and placed under shelter at night, for two reasons. First, if left spread out in the damp night air they may sour or mold and will certainly take much longer to dry. In the second place, moisture is "sweated out" while in a shelter protected from rain, for, or dew. The fish are stacked in two rows to each layer with the tails to the center and the napes to the edge of the rack, skin side down, except for the top layer, on a low slatwork rack about 6 inches high. The stack is covered with matting if there is any doubt that the shed is not dry enough. Weights are placed on top of the stack to press out additional moisture. Additional weights are added each night as drying progresses.

The time required for drying depends on weather conditions and the size of the fish to be cured. It is, therefore, not advisable to at-

Figure 17.—Drying salt fish in warm climates.

tempt to give the exact length of time required for air drying. In the experimental work conducted by the U.S. Bureau of Fisheries (now the Fish and Wildlife Service), 45 hours of air drying, or about 5 days in total time, was a fair average for the drying period. Therefore, due to the many variants, such as changes in the weather conditions and size of the fish, the time required for drying must be determined by the curer through practical packing operations. The fish may be considered cured when no moisture is apparent on the surface, and a thumb pressed into the thick part of the flesh makes no impression. The fish must be dried down to a moisture content of not more than 25 percent.

Preventing Insect Infestation During Drying

If the weather is still, especially during the first part of the cure, a trench should be dug along each of the two long sides of the drying roof. Fires are built in these pits, which are then smothered to form a smudge with green coconut-leaf ribs or husks, or branches from aromatic shrubs. This drives away flies which would otherwise infest the fish flesh with maggots. If humid, still weather sets in when only a few hours of drying are needed to complete the cure, or if the fish have been pressed and repressed in the kench (stack) for several days and no change occurs in the weather, so that they can be air-dried, the following method of air drying may be used.

Heat Drying

Light, movable walls of matting, bamboo, or thatch are constructed and set in place to close in the roof over the drying racks. Low fires of materials such as dry-coconut husks or dry hardwood are kept going in the pits so that there is a clear fire without much smoke. Doors of the drying shed are left open in the direction of any air movement. The fish are spread out on the racks and turned regularly every two hours. The fires must not be allowed to blaze, and the temperature at the level where the fish are laid out should be less that 100° F., or the fish will be scalded. If a thermometer is not available, a good rough test is to stretch out a hand among the fish, and if the air feels distinctly warm on the hand, the temperature is too high.

While not so good a quality as if air dried naturally, the product will be acceptable and better than fish cured locally by the old methods.

Handling in Storage

Dried fish should be stored in wooden boxes lined with tin plate, metal foil; or, failing these, with wax paper. They should be inspected at regular intervals while in storage. If there has been a period of damp weather, and the dried fish show signs of moisture, they should be given a few hours of air drying. If signs of rust and mold appear, the fish should be scrubbed in a light salt brine containing some vinegar, then spread out to dry in the air for a day or two.

Chapter Twelve

Miscellaneous Brine Packed Fish

At one time brine salting was relatively unimportant in comparison with dry salting, but both methods are almost equally important today in the United States. Cod is the most important dry-salted product; salmon the most important brine-packed product.

There is some confusion as to a definition of brine salting. According to Tressler (1920), brine-salted fish are cured by holding in a concentrated salt solution in a watertight container, while dry-salted fish are packed in watertight containers with an excess of dry salt. However, fish cured by the latter method cannot always be considered dry salted, and while pickled and mild-cured salmon are cured by placing them in watertight containers with an excess of dry salt, this is merely to allow the brine to form naturally by the extraction of moisture from the flesh by the salt. These salmon are cured and afterward held in brine. They are in dry salt only in the first stages, before curing has begun. Conversely, alewives or river herring are placed in a saturated brine, but as soon as it has penetrated they are packed in dry salt and held in dry salt until consumed. A product made in this way should really be considered dry-salted.

It is believed that the following definition is more accurate. Brine-cured fish are fish cured in a brine solution, either formed naturally by extraction of moisture from the flesh by dry salt or manufactured from salt and water, and held in brine until consumed.

Whenever possible, brine should form naturally by packing the fish in dry salt. This is in accord with the phenomenon of osmosis discussed under the section on Principles of Fish Salting. Enough brine to cover the fish will usually form in about 12 hours or less. In warm weather the fish are more liable to spoil than if they had been placed immediately in manufactured brine. Packing fish in dry salt requires more skill than is necessary in the brine curing method. It seems simple enough to add an excess of salt, but if that excess is too great the fish will be burned. The coarseness of the salt, the temperature and humidity, the composition of the salt, and the quality of the product desired, must be considered.

Brine-packed fish can be freshened to a condition more like that of fresh fish than may dry-salted fish. The moisture content of brine-salted fish is much greater than that of dry-salted fish, so that loss in manufacture is less and consequently raw material coasts are lower. The disadvantage of brine salting is that the product is less convenient to transport and distribute. Brine-packed fish require special storage if they are to be held for more than a short time. Brine-salted fish are not nearly so durable as dry-salted fish and they spoil quickly in warm climates, while dry-salted fish cured to a moisture content of 30 percent or less, will remain in good condition over a much longer period. Dry salted fish are not so dependent on cooperage and storage as are brine-salted fish. In long-period storage, where special low temperature storage is available, fish stored under brine change much less than dry-salted fish, when both are held at the same temperature.

ALEWIVES OR RIVER HERRING

Alewives or river herring are of two species (*Pololobus pseudoharengus* and *P. aestivalis*). They are salted commercially in Maine, Massachusetts, Maryland, Virginia, and North Carolina. North Carolina and Virginia produced 5,742,000 pounds of salt river herring in 1940, or 81.4 percent of the total pack for that year. The pack of salt herring has decreased in recent years, not so much as other cured products, because of decrease in demand, but because the preparation of vinegar salt-cure herring has taken much of the supply of available raw material. A great quantity of salt herring is now marketed as corned rather than as tight pack herring, which was formerly the standard pack. The curing of river herring as conducted in the Chesapeake Bay area is regarded as representing the best commercial practice. This has been described fully by Jarvis and Hines (1943).

Catching, Handling, and Landing

River herring, or alewives, salted in the Chesapeake Bay area, are caught in fish traps, locally called pound nets. If the distance to the fishing grounds if not too great a portion of the catch is brought to the processing plants in the boats used to fish the traps. These fish reach the packing house on an average of 5 hours after they are caught. The remaining fish are brought in by run boats. The load of run boats or transporters includes fish obtained from nets at some distance from the saltery and the catches of a number of trap operators, each with a single net, who are dependent on these boats to handle their catch. The fish brought in by run oats, as a rule, have been out of the water from a minimum of about 12 hours to a maximum of 24 hours. Under ordinary weather conditions a salted product of good quality cannot be prepared if the herring have been held longer than this time before salting, though should the weather be unusually cool, between 40° and 50° F., the time in which the fish will remain in good condition may be extended to 36 hours.

River herring are bought by count. The price averaged five dollars per thousand in the years just before our entry into World War II. The fish are unloaded from the hold of the transporting boat by bucket hoists. A bucket will hold about 500 fish. An occasional bucket of fish is counted to establish the number held by the container when it is filled level with the brim and to determine the proportion of scrap fish. These are usually small menhaden or croakers and are discarded, with no payment to the fisherman.

Washing and Butchering

The bucket is emptied into a hopper leading to a washer, which consists of an inclined revolving, wire-mesh drum with a perforated pipe running along the axis. One inch angle irons are fixed at one or 2 foot intervals along the inner circumference of the drum to act as baffles. As the fish are carried through the drum the scales are removed by the tumbling action. Sprays of water from the perforated pipe wash the fish at the same time. The fish are carried by overhead conveyer from the washer to the cutting table. Details in cutting-table construction differ from plant to plant, but a typical table stands about four feet high, with a width of some 8 feet. It rises to a peak in the center with a 1-foot slope to the sides. The fish fall from the conveyer onto the peak of the table then slide down toward individual pens where they are cut.

Each workman is provided with a wire fish-basket and an enamel or galvanized water bucket. The herring is grasped around the back near the head, and the workman makes a single curving cut, starting at the back of the head just behind the gills and continuing down the belly to the vent (anal opening). This removes head, viscera, and thin ventral flesh in one operation. The fish is thrown into the wire basket. If the herring contains roe, this is separated from the viscera and dropped into the galvanized bucket. If the cutting is not done carefully, some viscera may be left in the body cavity or the roe may be damaged by cutting. A large part of the cut roe is lost in canning, since it drains away in washing. Cutting is done on a piecework basis, the workers being paid a certain price for each basket of cut fish or bucket of roe.

After cutting, the fish are thrown into washing tanks. A typical tank is made of 2-inch pine planks and is about 12 feet long, 6 feet wide, and 3 feet deep. Some tanks are constructed with the bottom inclined about 30 degrees to one side, and a false bottom of slats, spaced one-half inch apart, is placed in the tank at the top of the incline. Scales, viscera, and other waste material settle down in the deep angle of the bottom. Waste is drawn off with the water through an outside pipe or flood gate. This type of tank is good because it requires less time to clean than an ordinary flat-bottomed tank, and waste is more readily separated from the fish. The fish are stirred about in the water of the washing tank for about 10 minutes, making washing more thorough. The stirring is done mechanically in some plants, but hand labor is still used in a number of salteries.

Salting

When the herring are sufficiently washed, they are scooped out of the tank with dip nets into slat cars holding about 1200 fish and carried to the salting vats. In some salteries wheelbarrows with high sides are used for this purpose. Excess water drains away as the fish are carried to the vats. These vats are shallow wooden tanks built of 2-inch Virginia pine. They are usually 10 feet long, 6 feet wide, and 3 feet deep. Before any fish are put in, brine testing 100° salinometer is poured into each tank to a depth of about 4 inches. As each load of fish is emptied into the tank, additional salt is added, the exact amount varying with the condition of the fish, humidity, and temperature, but averaging about 700 pounds to a tank. When full a tank contains from 12,000 to 15,000 fish (about 4,000 lb.), and the brine should test from 95° to 98°

salinometer. The fish should be roused; that is, stirred about once each day while curing. After each rousing they are pressed down lightly in the tank and a thin layer of salt is scattered over the top. About 25 pounds of salt are added each time this is done.

Cut or clipped herring require about 9 days to cure. In fact a curing period of this length is the standard required under Virginia packing regulations. Roe herring cured with only the heads removed must be salted 14 to 15 days. Strength of the brine is checked daily and should test not less than 90° salinometer. If it falls below this level, additional salt should be added.

Sufficiency of cure is determined first by the odor, when the flesh is pulled from the backbone. The odor should be good, without suggestion of sourness. The flesh should be firm, but not excessively hard, the skin only slightly wrinkled and the blood under the backbone should be dried.

Packing

When cured, the fish are taken from the tanks and piled in heaps on the floor of the packing shed to drain. It is recommended that draining racks should be at least 6 inches above the floor. The herring should not be piled in heaps more than 18 inches high. The fish should be drained from 4 to 7 days. The herring are then weighed or counted and packed in tight, heavy barrels. The fish are filled into the barrels in wheel-like tiers; that is, radiating from the center, with the first layer packed backs down, and all other layers with backs up.

A layer of three-quarter ground salt is scattered over each layer of fish, using about 2 1/2 pounds to the layer. A properly packed barrel should contain 160 pounds of fish and 50 pounds of salt, with an average of 600 to 650 herring by count. The variation in count between barrels in any one lot must not be more than fifty. The cured fish should measure not less than 6 inches if headless, or clipped, while whole fish should not measure less than 7 1/2 inches.

The barrel head is pressed in and the container headed up. It is then marked with the packers name or brand, and sometimes with the approximate number of fish. No brine is added, so that this pack, while brine cured, goes into the product as a dry-salted product. This cure is known as tight pack. It is subject to oxidation after a few months of storage at ordinary temperature.

Corned Herring

The fish may be taken out of the tanks after they have been in brine from 12 to 48 hours. When this is done they are usually drained from 5 to 24 hours, although some packers do not drain the fish at all. The drained herrings are packed in bushel baskets with about 250 fish to the basket. As a rule no salt is added. The baskets of fish are sold to grocers, or hucksters who peddle them throughout the countryside. This product is called corned herring and when boiled or fried is a favorite springtime breakfast dish in the southeastern United States.

Much of the salted river-herring are given the full 9-day cure, after which they are hauled in bulk to Richmond where they are stored in brine, and held throughout the year in chill storage at 34° to 36° F. until sold to the retailer. Then the fish are drained for a few hours and packed in bushel baskets as described above. They are sold as corned herring, though they are more heavily salted than the corned herring described previously. This pack answers the demand of the retailer for a smaller package. It is brighter and more attractive in appearance than the tight pack which it may displace entirely. Corned herring will remain in good condition from 10 days to 3 weeks in the hands of the retailer, the time depending on the length of cure, and on the temperature.

Inspection and Grading

A large part of the herring cured in Virginia are packed under a voluntary inspection system administered by the Division of Markets of the State Department of Agriculture (Jarvis and Hines 1943). Under this system the packer agrees to maintain certain standards of pack. His fish are regularly examined by qualified inspectors while curing to insure that the process in being properly done. The inspectors keep complete daily records of temperature, salinity of brine, condition of fish, and any other factors which may affect the cure. The inspector is present when the cured fish are packed for shipment to insure compliance with the accepted standards. If the pack is satisfactory, a grade certificate is issued to that effect, and the inspector stamps each container with the official grade stamp. The grades are as follows:

> **Virginia No. 1 Herring**—shall consist of fish of one species, taken before spawning, fresh when curing begins, bright, free from rust at time of packing, of a sweet wholesome odor, thoroughly cured, well drained, with 80 percent of scales removed, and free from damage from

any cause. All clipped fish roe shall have the heads properly removed. All roe fish shall have a fairly developed roe.

All cut fish shall be properly cut, thoroughly cleaned, and washed before curing begins.

Virginia Selects—shall consist of fish that meet all the grade and packing requirements of Virginia No. 1 and in addition shall, when clipped, have all membrane and blood removed from under the main bone. All roe fish shall have a well developed roe.

Unclassified—shall consist of fish which are not graded and packed in conformity with the foregoing grades

This system of inspection has done much toward raising the quality of the Virginia pack and to establish it as a high grade of salt fish product.

COD

Brine Packing

On the Atlantic coast, brine packed cod (*Gadus morhua*) is prepared commercially in small quantities, for local distribution. This is done principally by shore fishermen in the State of Maine. On the Pacific coast, brine packed cod (*Gadus macrocephalus*) is packed in the Puget Sound area and in Alaska. In addition to the local market, some brine-packed cod is shipped to the Scandinavian-American trade of the Middle West. Commercial fishermen on both coasts prepare this pack for their own use; the fishermen themselves preferring these fish since they rehydrate more readily, and have a flavor and texture superior to dry-salted cod.

On the Atlantic coast, pollock (*Pollachius virens*) also are prepared by brine curing. Haddock (*Melanogrammus aeglefinus*) were once brine packed to a considerable extent but this use is only occasional today. Hake and cusk are rarely, if ever, brine-packed commercially. The method is the same for all groundfish.

Butchering and Salting

The fish are dressed, split, and washed as for dry-salting. Large fish are sometimes split into halves. When the washed fish have drained sufficiently, they are salted in butts or hogsheads as when preparing for dry-salting. If was formerly customary to call for a special curing salt such as Liverpool, Trapani, or Lunenburg. This is no longer true. Any good commercial salt may be used provided it has a medium-size grain; three-quarters ground salt is preferred.

A thin layer of salt is scattered on the bottom of the hogshead. Salt is rubbed into the flesh and the fish are packed into the hogshead flesh side up. A scattering of salt is thrown over each layer of fish taking particular care that salt is present where fish overlap each other. Each layer is packed at right angles to the preceding one. The top layer of fish is packed skin side up, and a heavier layer of salt sprinkled on top. The top may be weighted down to keep the pieces of fish submerged after the brine forms. The top should be covered with brine in 24 hours of less, but at least a week will be required for the fish to be struck through or cured. An experienced curer can determine this by the appearance of the flesh. Salt is used in the proportion of 35 pounds to 100 pounds of fish (Jarvis 1944b).

Repacking

The cod are allowed to remain in brine until required for sale. They are then removed and repacked in barrels holding 200 pounds, half barrels of 100 pounds, and kits or tubs holding 25 or 50 pounds. Each piece of fish is scrubbed as it is repacked in the container and a small amount of salt is scattered on each layer of fish, using 15 to 20 pounds of salt per 200 pound barrel. After a barrel is headed, it is filled through the bunghole with brine testing at least 90° salinometer. The product ships quite well, and does not require special handling in storage when the barrels are tight and the brine does not leak out.

Some of the smaller operators pack the cod directly in shipping containers after splitting, adding salt in the proportion given above. Brine forms naturally. When the cure is complete and the fish have settled down in the barrel, a few more fish of the same day's cure are added to the top. The barrel is then headed up. Some curers also drain off the brine formed in curing, refilling with fresh 100° salinometer brine.

Pickled Cod European Method

It has been reported that brine-salted cod of higher quality than that prepared in the United States has been packed in some of the European countries. Brine-salted cod (or pickled, as it is called in the British Isles) is cured in greater amounts and is more popular than in the United States. It is packed commercially in Holland, Scotland, Sweden, Belgium, and France. The choicest pickled cod are reported to be prepared in Holland. The method given here is a summary of the best European commercial practice. The method used varies in Holland, Scotland, and Sweden, only in such points as the type of salt used or in the minor details of splitting.

When possible, the cod are bled as soon as caught. They are headed and gutted at once when received at the curing plant and put into clean fresh water. The fish are then washed and cleaned thoroughly. After draining, the fish are ready for splitting, which is considered to be a very important factor in preparation. The recommended practice is as follows:

The gutted fish is laid on the butchering table with its tail toward the splitter, who takes hold of the upper napes of the cod with his left hand, enters the knife at the vent and draws it down to the tail above the bone. He then gives the fish a half-turn, its tail falls outward and the shoulders swing inwards till the back is turned towards him. He raises the napes with his left hand, makes a cut just above the end of the backbone at the head end, and with the knife blade at a downward angle, carefully splits the fish from the bone, leaving as little flesh on the bone as possible. The split fish is given another half turn so that the head is toward and the tail away from the splitter. He steadies the fish with his left hand and carefully runs the knife down under the backbone, with the edge of the blade at a slight upward angle, separating the backbone from the flesh. The backbone is cut off a short distance above the tail. the cut is usually made through two vertebrae at an angle. The outer ends of the rib-bones, if still adhering to the flesh, should be carefully cut, not torn away. A slash is made just under the remaining backbone, so that it may be bled properly, and the salt may penetrate more readily.

The next step is cleaning. The black skin or belly membrane is removed and any bits of viscera, coagulated blood, or ragged edges of flesh or skin should be carefully cut away. The fish is then washed thoroughly in clean water, using a scrubbing brush both inside and out. The washed fish are piled in small stacks, skin side up. The stacks are covered with boards and weighted down lightly. The fish are left in the stacks about two hours. This is considered more effective than ordinary draining and also presses out diffused blood not removed in washing.

The cod are then salted into butts or tierces. A scattering of salt is thrown on the bottom. The first layer of fish is laid in skin side down. A thin layer of salt is then added. It must be spread evenly and the fish must be completely covered. The second layer of fish is packed in at the same angle as the first, the flesh side down. The third layer of fish is packed at right angles to the first two, with the flesh side up. The fourth layer of fish is packed at the same angle as the third with the flesh side down. This is done so that the flesh in salting will not be discolored by the skin pigment of the fish in the layer above. The amount of salt to be used depends to some extent on the season, the size and thickness of the fish, and the length of time they are to be held before marketing. As a rule an average of 84 pounds of salt are used to 250 pounds of fish.

The cod are left in the curing vats for 48 hours, then taken out of the brine and washed carefully to remove slime or excess salt. After washing they are laid out on a packing table, flesh side down. Any wrinkles are smoothed out carefully. Anal fins should be cut away neatly and dangling tags or ragged edges of flesh or skin should be pared off, to give a neater appearance.

As the fish are trimmed they should be graded into two or three different sizes, each grade to be packed into separate barrels. The number of fish contained should be marked on each barrel.

Figure 18.—Brine-packed cod showing arrangement in the barrel.

The barrels should be wet thoroughly before use. A scattering of salt is thrown on the bottom. In packing the fish the packer must lift and handle them carefully so as not to damage them. Duthie (1911) reports that the best way is to grasp the tail of the fish with the right hand and its shoulder with the left, the skin of the fish being downward. The fish will thus fall in a partial fold and allow for its being put inside the barrel easily. The bone of the fish should be laid next the side of the barrel. Two medium sized fish will make a tier laid head and tail alternately (fig. 18). The layers are laid in flesh side up except for the top layer, and at right angles to each other. If the fish are to be used at once no salt is needed between layers, but as the length of time required for shipping and retailing is uncertain, a sprinkling of salt is advisable between layers. This amounts to about 15 pounds to the barrel.

When completely filled, the barrels are headed up and a bunghole bored in the bilge of each one. The barrels are filled with fresh strained 95° to 100° salinometer brine through the bungholes. If held in storage the barrels should be inspected regularly to see that they are full of brine. If held for longer than 6 weeks the fish should be repacked before shipping. If necessary they should be washed and trimmed again and the barrels filled with fresh brine.

Cod Tongues

Cod tongues are considered a delicacy, especially by New Englanders, Nova Scotians, Newfoundlanders, and Scandinavian-Americans. While most of the preparation and preservation of cod tongues for home use, they are also cured commercially. The commercial production of salted cod tongues in the United States was 44,000 pounds in 1943.

Cod tongues are saved if the butchering crew is not too busy. In the vessel fishery one of the dressing gang usually cuts out the tongues, while at shore stations a boy is hired to do the work. The tongue is attached to the lower jaw, and when cut includes all that part of the jaw lying inside the jawbone. To cut out the tongue the workman takes hold of the fish by the back of the head, using the bony eye-sockets for finger holds. As he lifts the fish by the head its mouth usually falls open, then with his other hand he cuts the tongue loose on the sides with a sharp knife, following with a cut along the curving bone forming the back part of the lower jaw. The tongue is hanging by a thin strip at the forward of the lower jaw from whence it is torn loose by hand. The tongues are washed thoroughly in clean sea water and trimmed of any loose skin or fragments.

After draining, the tongues are mixed with salt in barrels following the proportions of 25 pounds of salt to 100 pounds of tongues. The tongues must not be oversalted or they will soon become hard and have an acrid flavor. Brine is formed by the extraction of moisture from the tongues by the salt. The tongues require an average of 10 days to become struck through, that is, for the salt to penetrate completely. When the tongues are sufficiently cured they are rinsed in light brine and repacked in barrels, with 200 pounds of tongues, net weight, to the barrel. The containers are headed up, after which they are filled with 100° salinometer brine. Wholesalers or jobbers again repack the tongues for sale to the retailers, using half barrels holding 100 pounds of tongues; kegs, with net contents of 50 pounds; pails holding 25 pounds; and kits or small pails holding about 10 pounds.

For table use, the tongues are soaked for several hours in fresh water to freshen, wiped dry, rolled in beaten egg and cracker crumbs, and fried. The salt tongues cannot be held for more than a year even if well cured. The tongues gradually harden after about 9 months in salt, becoming inedible. (Cobb 1926).

LAKEFISH

The lake herring (*Leucichthys artedi*) is the only fresh-water fish cured in large quantities by salting in the United States. The production was 5,037,000 pounds in 1940. This is about 32 percent of the catch for the year as there is a loss of approximately 30 percent in salting. Lake herring are salted along the shores of Lake Michigan, Lake Huron, and Lake Superior. According to the United States Tariff Commission (1927), the bulk of the lake salted herring salted in the Lake Michigan and Lake Superior areas is cured by the fishermen. The salt fish companies as a rule supply the fishermen with barrels, salt, and any other necessary materials and buy the salt fish as cured, for final preparation. On Lake Huron the fish companies usually do their own curing. Some of the Lake Michigan fish companies also do their own curing. The most important types of gear used in catching lake herring are trap and pound nets in Lake Huron and gill nets in Lake Michigan.

Lake herring for salting are brought into the fish house as soon as possible. During the summer months ice must be used on the fishing vessels to keep the fish in good condition. The herring are dressed and cleaned as soon as they are brought ashore. The head is removed, then a cut is made down the belly to the vent and the fish are eviscerated. The loss of weight in dressing averages 18 percent. The fish are thrown into a box as they are cleaned. When the box is filled it is emptied into a wash trough of fresh water and washed thoroughly to remove any blood, slime, or other offal.

When sufficiently washed, the lake herring are removed from the washing tank by a dip net and emptied onto the center of a salting table so constructed that the water will drain away from the pile of fish in the center. The common procedure in salting is as follows: Five fish are picked up and rubbed with salt. Particular attention is paid to the belly cavity. The fish are then packed together, back to belly. The salted fish are placed to one side and the process is repeated with another 5 fish. When 25 fish have accumulated, they are packed belly side up in a half barrel (capacity 125 lb. of fish). The method of packing is the same for sea herring (split-cure). A thin layer of salt is thrown over each layer of fish.

The salted herring are packed to a level of 4 inches above the barrel top, as there is considerable shrinkage in salting. The brine is allowed to form naturally. After standing for 12 to 24 hours, sufficient moisture has been extracted so that the barrel may be headed. From 7 to 10 days is required for the salt to penetrate completely, or "strike through" the fish. The herring are now known as "slime" fish. They may be repacked within a few days after the original salting, or held in this first cure for a week or more. If the herring are to be held, the barrels are headed, filled with 100° salinometer brine, and placed in chill storage. The loss of weight in the first salting is 9 percent.

When the lake herring are to be repacked they are removed from chill storage and the barrels are emptied onto a draining table 6 feet square, with a capacity of 5 barrels of fish. When the original pickle has drained away, the herring are transferred to a weighing table in the amounts needed to fill the shipping container.

> "The weighing table and the draining table, the surfaces of which are on a level are placed side by side. Rapid handling of fish is facilitated by having the weigher concentrate his attention on the one operation. The weighed fish are passed from the 'weigher' to the 'packer' by a revolving table, fitted with compartments. As a batch of fish is weighed it is thrown into one of the compartments, from which the packer on the opposite side removes it as the rotating table brings it within his reach (Tariff Commission 1927)."

With the exception of the top one, all the layers are packed backs down. Before the tops are fixed in place a small amount of salt (averaging 3 lb. per 100 lb. of fish) is added and the containers are filled with brine testing 100° salinometer. The salt lake herring are packed into kits or pails holding 5 to 25 pounds and in barrels with a net weight of 300 to 100 pounds. The loss of weight in repacking averages five percent, being less in winter and more in summer. The total loss of weight averages 32 percent of the original weight of the fish. The product should be held in chill storage if it does not go into retail consumption within a month or six weeks.

MACKEREL

Production and Imports

Brine-salted Atlantic mackerel (Scomber scombrus) also known as "Boston" mackerel, was at one time one of the most important fishery products in the United States. The peak of production in the industry was in 1830 when 449,950 barrels (99,990,000 pounds) were produced having a value of $1,862,000. The industry had its largest financial return in 1864 when the 324,454 barrel yield (64,990,800 pounds) had a value of $7,000,000 to the primary producer. Of the 131,939,255 pounds of mackerel taken in the United States in 1880, 80 percent was salted. In the 1880's, the 90's, and the early years of this century, salt mackerel was sold in almost every grocery store throughout the eastern States and the Middle West.

In 1940, which may be considered the last prewar year of normal production, the amount of Atlantic mackerel salted was 2,158,000 pounds, with a value of $135,669, while the total catch of Atlantic mackerel that year was 40,631,000 pounds. Split salt-mackerel was formerly the standard product. Two-thirds of the production now consists of salt-mackerel fillets. At one time the salting of mackerel was general throughout the New England States. It is now confined entirely to Massachusetts, with Gloucester as the most important production center.

Salt mackerel is also imported into the United States from Canada and several European countries, principally from Eire and Norway. Some mackerel is salted in Great Britain, France, and the Netherlands. Mackerel imported into the United States is cured, packed and graded by the American method, as the United States has been the principal market for the past hundred years. Imports of salt mackerel have decreased considerably. They amounted to 12,071,000 pounds in 1927 and 3,180,000 pounds in 1940.

Method of Catch

Mackerel which are to be salted are caught almost entirely by purse seine. Trap caught fish may be used at times. Formerly the bulk of the fish which were to be salted were cured at sea. The vessel cure was the typical method and is the one described by Stephenson (1899) and Tressler (1923). Today practically the entire pack is prepared ashore, utilizing the surplus from the fresh-fish market. Salting is done aboard the fishing vessel today only to save small catches which do not justify a run to port.

Butchering

The method of preparation is essentially the same whether at sea or ashore. Only fresh mackerel should be used for salting. Any attempt to "save" mackerel which is not quite fresh will result in an inferior product with a very rapid rate of deterioration. Since mackerel spoil readily, they should be handled as rapidly as possible after catching. If

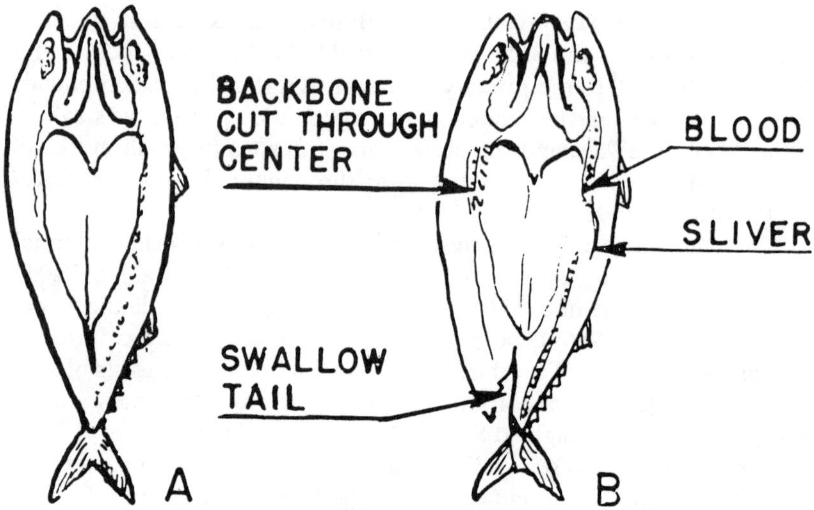

Figure 19.—Splitting mackerel for salting:
(A) Properly; (B) Improperly split fish.

not split immediately, the mackerel should be iced, protected against the sun in warm weather, and should not be piled in heaps.

A dressing-gang usually consists of three men, one splitter and two "gibbers," who eviscerate and clean the split fish. In Norway, the gills are sometimes removed before splitting. As the appearance of the fish when cured depends a great deal on the cutting, splitting must be done accurately. Sharp knives must be used for a clean cut. The mackerel are split down the back, as close to the backbone as possible., and on the left side. Each fish is laid on the splitting board with its head away from the splitter who begins at the head and runs his knife down the backbone to the tail, cutting so that the mackerel will be flat and open in a single piece after the viscera are removed (fig.19). For protection, and to prevent the mackerel from slipping the splitter usually wears a cotton or woolen glove on his left hand, which is holding the fish.

As each mackerel is split it is tossed to a gibber who opens the fish with a jerk, causing it to break along the lower end of the ribs, if it is fat, making a crease along each side. The gibber then removes the viscera and gills. If a fish is not fat he often "plows" or, as it is sometimes called, "reams" the mackerel. This consists of making a shallow cut in the abdominal cavity near each side of the backbone in imitation of the natural breaks occurring in fat mackerel. At one time this operation was done with the thumbnail but a special knife has been used for years.

Washing

Clean salt water is often used for washing. Tubs of clean water should be standing near the gibbers' end of the splitting bench, and each mackerel is dropped in the water as it is gibbed, with the fish open, and flesh side down. Ragged or soft fish should be discarded when splitting. Some curers prefer to leave the reaming or plowing of mackerel until the fish have been in water for some time. In washing the fish, care must be taken to remove all traces of blood, especially around the backbone. Some curers scrub out the blood with small brushes, but most of them clean out blood with the hands only.

The mackerel is a tender fish that requires careful handling during all stages of preparation to keep the flesh from being broken. In lifting split mackerel before it is completely cured, that is, salted, it is recommended that the hand should be placed below the fish which can then be lifted without being gripped. If handled at all roughly, the flesh has a ragged appearance when cured. Any protruding rib or other small bones must be cut off or plucked out with forceps.

As each fish is washed, it should be dropped into another tub of clean water. If possible there should be a steady flow of water through this tub. If it is necessary to use standing water, the water in the tubs should be changed three times for each batch of fish. The first water should be emptied at the end of 15 minutes, as the blood flowing from the fish soon colors the water, and if left in the tub for any length of time would discolor the flesh of the fish. The fish can be left a little longer in the second water, and can stay in the third water for about an hour. According to Stevenson (1899), mackerel were kept in the wash tubs to soak until all the mackerel were split, so that some mackerel may have been in the water for as much as 6 or 8 hours, or even longer. This is now considered poor practice, as the mackerel become too soft and are subject to breakage. The mackerel should not soak for more than 2 hours as a maximum. After soaking, the fish are rinsed by throwing buckets of water over them.

Packing

Only new barrels should be used in packing mackerel. Some curers try to economize by using second hand fish barrels, but these containers are apt to give an unpleasant foreign flavor to the fish. The Norwegians generally use barrels made of birch. American and Irish curers find barrels made of spruce satisfactory in packing mackerel. The mackerel barrel usually has an iron hoop on each end, with three or four wooden hoops made out of green withes.

The barrels should first be soaked in water to make sure that they are tight, for if leaky, or if the staves soak up too much brine, the fish will be exposed, causing "rusting" or oxidation, and spoilage.

After the mackerel have been rinsed and drained, they are placed, a few at a time, in a shallow tub partly filled with dry salt. The flesh side of each fish is rubbed in the salt, and they are picked up with as much salt as will cling to the body. A thin layer of salt is scattered on the bottom of the barrel. Then the first layer of fish is packed with the skin side down. The mackerel are packed in circular fashion, with the tails toward the center of the barrel. If there is a hollow space in the center, where the tails meet, one or two mackerel are laid across this space, to keep the layer level. A thin layer of salt is then scattered over the layer of fish. It is most important that every portion of the surface of the mackerel come in contact with the salt. After the first two or three layers, the mackerel are usually packed flesh-side down.

Salt Requirements

For salting mackerel, Americans and Irish generally use Liverpool No. 2, a finely ground fish salt, while Norwegians prefer Trapani salt. Coarse-grained fish salts such as Cadiz should not be used, since they pit the surface of the fish and give them a ragged appearance. If the mackerel are fat, 100 pounds of salt will be required for 300 pounds of fish. If fat fish are not heavily salted they will stick together in curing, making it impossible to separate without tearing. This would make them unsalable. If the mackerel are not fat, the proportion of one part of salt to four parts of fish will be satisfactory.

The barrel should be filled up to the croze, that is, the groove into which the head fits, with a heavier layer of salt than usual scattered on top. The barrel is then headed up, and a bunghole is bored in through the side. Strong brine (90° to 100°) salinometer is poured in through a funnel until the barrel is full. The bung is then driven in and the barrel laid on its side in a cool warehouse to complete the cure.

Curing and Repacking

From 10 to 12 days will be required to cure the mackerel, that is, to obtain the maximum penetration of salt. The barrels of salt mackerel are not rehandled but remain in storage until required for market. While in storage the barrels should be examined regularly for leakage and to see that they are completely filled with strong brine. The barrels are coopered, and additional brine is added as soon as any loss is detected. If the fish are allowed to become dry, they soon become soft and discolored, with a strong, unpleasant taste.

The mackerel are repacked carefully before shipping. The tops of the barrels are taken out, and the brine is drawn off and discarded. Then the fish are emptied, several barrels at a time, into a bin known as a "culling crib." It is a box of planed wood, about 5 feet long, 3 feet wide, and 8 to 10 inches deep, with a slat bottom, placed on legs 3 feet high. Here the mackerel are culled, or separated, into the commonly recognized grades. From the culling cribs the fish are thrown into weighing tubs. These tubs hold about 100 pounds each, and have network bottoms or a one-inch hole in the bottom for draining. When a tub is filled, it is weighed and taken to the packing cribs. The packing cribs are something like the culling cribs. The usual dimensions are 38 inches long, 26 inches wide, and 14 inches deep. The mackerel are packed into barrels, kegs, kits, or other containers from the packing crib.

Each grade of mackerel is kept separate and packed into separate containers. In repacking, the fish are filled in as in curing, with a layer of salt on the bottom and a handful of salt over each layer. The first two layers go in with the skin side down, all succeeding layers with the skin up. About 36 pounds of salt are used in repacking each barrel, which should contain 200 pounds of cured mackerel, exclusive of salt and brine. After being filled, each barrel is headed up and moved to a convenient place in the warehouse, where it is filled with strong brine. The strength of the brine should be at least 95° salinometer. After a barrel is filled with brine, the bunghole is plugged and the barrel is set on end to be branded. The brand includes the name of the firm curing the mackerel, the net weight, number of fish in the barrel, the year, and sometimes the month when the fish were packed.

Quality and Grades

There is much variance in the quality of salt mackerel taken at different seasons of the year. The early spring catch is generally very lean and shrinks considerably when salted. The fish are fatter as the season advances, and those taken in the fall usually improve in weight after being placed in brine. Full grown fresh mackerel measure 17 or 18 inches in length. The average length, however, is about 12 inches and the weight a trifle less than a pound. Salted mackerel measure less due to shrinkage in salting.

The grades usually recognized in commercial practice may be defined as follows: Best quality, not mutilated, measuring, when split, not less than 13 inches from the extremity of the head to the fork of the tail, and free from taint, rust, or damage, are branded as No. 1. Mackerel of the same quality except that they measure as short as 11 inches are classed as No. 2. Second quality mackerel but free from taint or damage and not less than 13 inches in length are graded as No. 3, large. Mackerel shorter than 11 inches, yet free from taint or damage are called either No. 3, small or No. 4. These grades have been put into the statutes of several of the New England states with little difference in specifications among states.

Salt Fillets

Salted mackerel fillets were first cured commercially in 1928. The salted fillets are now much more important in the salt mackerel trade than whole split mackerel. Fillets are prepared by cutting the head off just below the pectoral fins. A sharp knife is inserted just above the backbone at the head end. Then, keeping the blade at a slight downward angle, a sweep of the knife removes the upper side of the fish. The knife is then inserted just below the backbone and with the blade held at a slight upward angle, that backbone is removed, leaving two boneless pieces of fish. The thin edges of belly flesh are trimmed away.

The fillets are soaked in clean water, as in curing split mackerel, but are removed in a shorter time since the blood is more easily soaked out with the backbone cut away. The curing process differs little from that for split salt mackerel. The fillets are packed in barrels of the same size as used for split mackerel. The amount of salt used, averaging 30 pounds per barrel, is somewhat less than for split salt mackerel. The barrels are headed up, filled with brine, and laid away to cure. The loss in splitting and curing averages 50 percent.

The fillets are held under the same conditions as split salt mackerel until required for market. They are taken to culling cribs where the fillets are sorted for size, and any defective fillets are discarded. The fillets are repacked in kegs holding 100 pounds and tubs or kits with net contents of 10 to 50 pounds. The 10 pound kit is a favorite package. Little or no loss from shrinkage occurs in the repacking process.

MULLET

The striped mullet (*Mugil cephalus*) is the most important food fish of the southern coastal states. It is found from North Carolina to Mexico. This fish, or related species, is also found in Mexico and in Central and South America. The United States catch of mullet amounts to more than 30,000,000 pounds annually. Most of it is used fresh. From one to 2 million pounds are salted annually, mostly in North Carolina and Florida. Dry-salting is the method for curing mullet on the Gulf coast of Florida, but on the Atlantic Coast of Florida and North Carolina, brine-salting is used. Brine salting was formerly more extensively employed, but as the result of a change in the food preference of the southern population, much of the market has been lost. Brine-salting is usually considered necessary with fatty fish because the fat is less apt to become rancid than in the dry-salting process. Attempts at brine-curing by curers who were both careless and ignorant of the proper methods, resulted in a very poor product and this was another important cause of the decline in the market. With improvement in preparation of the product, the trade could be reestablished.

Although the southern coastal states offer good markets for fish when it is available, the facilities for handling fresh fish are limited, with the result that there are stretches of coast where the mullet is almost unused. If a brine-salted mullet of good quality were prepared and marketed, the food supply of the country would be increased, and an appetizing protein food would become available to low-income groups in the southern states. The method described here is representative of good commercial practice.

Mullet are taken commercially by haul seines, and gill, trammel, pound, and cast nets. The runaround gill net is the most important type of gear accounting for 69 percent of the total catch in 1940. The haul seine is second in importance, credited with 15 percent of the catch.

Splitting, Cleaning, and Washing

Speed in handling is essential to a successful cure. Mullet must be split and cleaned as soon as landed. For a satisfactory product the maximum time allowable after catching is 6 hours. The fish are first rinsed thoroughly in clean salt water to remove slime and sand. The heads of all but the smallest are removed by cutting along the forward edge of the napes, or collarbone, which should be left in. The fish are then split down the back from head to tail, so that they will lay flat in one piece. To avoid a ragged appearance the knife should not cut entirely through the body in the tail section. A cut is made under the backbone also, to aid in penetration of the salt. If the mullet weigh more than a pound and a half each, about three-fifths of the backbone may be cut away, leaving only the tail section in the flesh. On larger mullet, the flesh is scored to a depth of about 1/2 inch in lines parallel to the backbone. Often both medium and larger sized mullet are filleted before salting, especially if they are to be packed in kits or tubs.

After splitting, the mullet are eviscerated. The black membrane lining the belly cavity should be removed. This is best done by scrubbing with a piece of coarse canvas or sacking. Thorough cleaning helps to remove blood and bits of membrane. The fish are then trimmed of any ragged edges, and washed in clean sea water or in a 50° salinometer brine, the latter being preferable. It is also good practice to soak the fish in brine from 30 minutes to an hour to remove diffused blood from the flesh.

Salting

After soaking, the mullet are heaped in a pile flesh side down, to drain for about 10 minutes. The fish are salted in vats holding about 400 pounds, though fountain syrup barrels or tierces may be used. Barrels that have held meat, butter, or pickles, are likely to give an "off" flavor to the fish and should not be used. A thin layer of salt is scattered on the bottom of the butt or salting tank. The fish are then taken singly and dredged in salt, rubbing it well into the flesh, especially into the cuts made in the surface. The salt used is of packers' fine or dairy-fine grade. A somewhat coarser salt, three quarters ground is sometimes used. Some packers use half-ground salt, which is not good, because the coarse crystals puncture the flesh and are slower to dissolve. Fine salt gives better results, although more care and accuracy are required in salting. As a rule, one part of salt should be used to three parts of fish. If the mullet are large and fat, the salt may be increased to about one part of salt to two parts of fish. An excess of salt will "burn" the flesh, giving it an acrid, unpleasant flavor.

The fish are laid in the butts or large barrels, skin-side down. A sprinkling of salt is scattered over each layer. Another layer of fish is laid at right angles to the first. The last layer is packed skin-side up and covered with a heavier layer of salt. The top is usually weighted down to keep the fish covered with the brine or pickle.

Sufficient brine to cover the mullet should form in from 12 to 24 hours. The fish should remain in the brine until they are "struck". This varies from 72 hours for the smaller fish to 10 days for the largest. The average time is a week. In warmer weather the time is reduced. The texture of the flesh determines when the fish are struck; the flesh should feel decidedly firm when pressed between thumb and forefinger, and the depressions should disappear slowly (Jarvis 1945a).

Repacking and Storage

When the mullet are struck they must be repacked. They are taken out of the salting butts, and scrubbed in clear brine to remove slime, undissolved salt, and other debris. They are graded as to size, thickness, and condition, and allowed to drain for one or two hours before packing in the final container. A barrel holding 100 pounds is the standard container for salt mullet, though tubs in sizes holding from 20 to 50 pounds are sometimes used. Mullet fillets are always packed in tubs.

A scattering of salt is thrown on the bottom of the container and the fish are laid in flesh side up, with the thick side against the side of the container. The mullet must be packed evenly and smoothly. A scattering of salt is thrown over the layer of fish and a second layer is packed at right angles to the first. This is continued until the container is filled, with the last layer packed skin side up. The amount of salt used in repacking should be about 10 pounds to 100 pounds of fish.

When filled, the containers are headed up and fresh 90° to 95° salinometer brine is added through the bunghole, until no more can be absorbed. The bung is then driven in and the barrel placed in chill storage at a temperature of approximately 40° F. Some of the fillets and larger choice fish are repacked in brine tanks in chill storage, and are not packed in the final container until ordered by the retailer. While in storage the containers must be inspected for leaks at regular intervals. Even when a container does not leak, there is a loss of brine from absorption by the fish; therefore, brine should be added at regular intervals. Oxidation or rusting will occur if the mullet are exposed above the surface of the brine.

In storage in the southern states at average room temperature, the salt mullet is reported to be at its best in from 1 to 6 weeks after packing; the maximum storage period is less than 6 months. It is believed that mullet should not be used after 3 months of storage at room temperature. The fish held in chill storage remain in excellent condition for a considerably longer period.

SABLEFISH OR BLACK COD

The sablefish or black cod (*Anoplopoma fimbria*) is found from Southern California to Alaska, with the greater part of the catch made off the coasts of Alaska, British Columbia, and Washington. They are caught mostly by long line, incidental to the halibut fishery, though with the development of the Pacific otter trawl fishery, increasing amounts are taken by the latter type of gear. Sablefish are caught all year round but the smallest catches are made when the halibut boats are not operating. Production of salted sablefish has remained at a level of 80,000 to 100,000 pounds annually for the past 20 years.

The sablefish is one of our best food fishes, and makes an excellent brine-salted product. It is one of the richest and fattest of American fishes with a firm, white, flaky flesh which has an appetizing flavor. In spite of its high fat content the salted sablefish does not oxidize readily when properly prepared and handled.

Sablefish are cured as follows: The head is cut off, retaining the napes (collarbone). A cut is made down the belly to the vent and the fish is eviscerated. Then it is split in two sides, using much the same method as when butchering for mild-cured salmon. Some packers split down the back, leaving the belly whole, as in preparing hard-salt salmon. The backbone, however, is always removed. For convenience in salting and repacking the first method is best for large fish, while the second gives best results with small. Sablefish are washed thoroughly after splitting, then allowed to drain until all surplus moisture is removed.

The sablefish are salted in mild-cure tierces or salting vats. A thin layer of salt is scattered on the bottom. The sides of fish are dredged in mild-cure or three-quarter ground- salt, then packed in the vat in a level layer, flesh side up. Salt is scattered over the layer of fish until no flesh is exposed. A second layer of fish is then packed at right angles to the first layer followed by a layer of salt. This is continued until all the sablefish are packed in salt with the last layer skin side up.

An average of 40 pounds of salt is used per 100 pounds of fish. Because of the fatness of the fish it is necessary to use more salt than with other species. The top of the vat is weighted down and the brine is formed as salt extracts moisture from the fish. It requires approximately three weeks for the salt to strike through, a longer time than for most other varieties.

When sablefish are thoroughly cured they are repacked for shipment. The sides are removed from the salting vat and scrubbed in brine to remove slime, dirt, and excess salt. After the fish have drained sufficiently they are packed in barrels to a net weight of 200 pounds. Repacking follows the same method as for hard-salt salmon or brine-cured cod. A little fresh salt is scattered over each layer, using about 15 pounds to the barrel. The barrel is headed up and it is filled through the bunghole with fresh brine testing 95° to 100° salinometer. Salted sablefish is marketed mostly through the Middle West, where it is a favorite product among Scandinavian-Americans.

Chapter Thirteen

Herring Salting

SPECIES CURED AND EXTENT OF HERRING FISHERY

Two species of herring are taken in North America. One is found on the Pacific Coast (*Clupea pallasii*), the other on the Atlantic coast (*Clupea harengus*). These fish are classified as separate species because of minor anatomical differences, which are indistinguishable to the layman.

The herring are confined to the Sub-Arctic and North Temperate zones. On the Atlantic Coast of North America they range from Cape Hatteras to Labrador, but the commercial fishery does not extend south of New Jersey, and the only really intensive fishery is on the Maine coast in the United States and the Provinces of Nova Scotia, New Brunswick, and Newfoundland in Canada. On the European coast the range of herring extends from northern Norway to the coast of southern France. The most intensive fisheries are off the coast of Norway, Scotland, and in the North Sea. The ranges of the Pacific herring is from San Diego in the south, to Nome in the Bering Sea, and to Kamchatka and northern Japan in the eastern Pacific. The only important United States herring fishery in the Pacific is on the coast of Alaska. British Columbia also has a very extensive herring fishery.

On a world-wide basis, the herring is the most important preserved fishery product. The largest amount is preserved by brine-curing, with smoking in second place, and canning, third. In Europe, the greatest production of cured herring is in Norway, with the United Kingdom, Netherlands, Iceland, Sweden, and Denmark next in order. In North America, Newfoundland is the most important producer, followed by Nova Scotia, and Maine. On the Pacific coast, Alaska ranks first in herring curing, with British Columbia second.

DEVELOPMENT OF HERRING CURING

The preparation of brine-cured herring has been an important industry since the Middle Ages. The first authentic writings dealing with herring-curing date from the twelfth century. Cured herring was one of the principal articles traded on the continent of Europe by England. The herring fisheries were the cause of several wars in the Baltic between some of the Hanseatic cities and the various Baltic states claiming grounds fished by the Hanseatic towns. The wars between England and Holland are held to be caused in general by their rivalry in trade and in the acquisition of colonial possessions, but the immediate cause was a struggle for control of the herring fishery on the East Anglian coast.

The method of herring curing is considered to have been crude until the time of William Beuckels, a fish merchant of Biervliet, in Flanders, who, during the fourteenth century, greatly improved the methods in use. This new development laid the foundation for the great wealth acquired later by the Netherlands from the fish curing business. Beuckels died in 1397, and his work was later considered so valuable that a monument to his memory was erected in his native village by Charles V.

The first mention we have of pickled herring in America is by Josselyn, in the seventeenth century, who, in his Chronological Observations of America, state: "We used to qualify a pickled herring by boiling him in milk." It is believed, however, that the pickling of herring was carried on by the earliest settlers of America, and possibly by the fishermen who came to these shores from Europe even before the first settlements were made, since the herring were readily caught in shore waters and herring curing was even then the most important fishery industry in Europe.

The first recorded commercial pack of brine-cured herring on the Pacific coast of North America was made at Fort Wrangell, Alaska, in

1878. Development of the industry was slow, the pack varying between 2,000 and 4,000 barrels, mostly "bloater stock" (for smoking as bloaters), until World War I, Scotch-cured herring was no longer available in the eastern United States market. It was thought that a herring-curing industry could be developed in Alaska to fill this demand, so in 1918 the government sent A.H. Klie, an expert herring curer, to Alaska, to introduce the Scotch style of cure. Production rapidly increased to 145,000 barrels in 1922. This is the record pack of brine-cured Alaska herring. The Alaska herring-curing industry is now at a very low level. The pack of 1911 was 4,651 barrels, and the pack of 1944 was reported as 4,159 barrels. This decline was accelerated during World War II, when Alaska became a war zone. It was evident, however, for several years previous. The decline is believed due to several factors, such as overfishing, high labor costs, high freight costs, foreign competition, and the irregular quality of the Alaska pack.

TYPE OF PRODUCT

A number of methods are used in brine-salted herring. It is almost impossible to list or discuss all the types of herring curing, but the six principal methods of brine-salting are: (1) round cure; (2) split cure; (3) Scotch cure; (4) Dutch cure; (5) Norwegian cure; and (6) matje cure. The Scotch cure is the most important commercial process used in the United States. It is a mild cure with limited keeping quality and is not suitable for some purposes because of the flavor given to it by the "blood pickle" employed in curing. Most of the miscellaneous types of brine-salted herring are given a much heavier cure and will, therefore, keep longer. As a rule, the largest herring obtainable are used for salting.

ROUND CURE

This is also known as the bulk cure. It is chiefly used to cure herring intended for smoking or spicing, although it is sometimes employed in preparing bait herring or for curing a few barrels of herring for local sale. On the Maine coast, the herring may be caught by weirs of seines. The season for large fish taken by weir usually runs from August to about November 15. Seine herring are caught offshore during the winter months. On the Alaska coast, in the Cook Inlet area, herring given the round cure are taken by gill nets during the winter months.

The fish are rarely if ever cured at sea, although this was formerly customary (Stevenson 1899). When the herring are loaded into the hold a little salt may be scattered on them if it is probable that the fishing boat will not reach port with a few hours, but in winter this is not considered necessary.

When landed, the herring are washed to remove scales, blood, slime, and trash, such as seaweed. Washing may be done by a hose, or in a drum (squirrel-cage) washer. After being drained for a few minutes, the herring are packed in wooden vats or tanks. These vary in size, but a vat 3 feet deep, 3 feet wide, and 10 feet long, is typical. Salt is mixed with the herring as they are placed in the vat. Experience and judgment are needed to determine the proper amount of salt required for curing. The quantity varies according to the size of the fish, their fatness, and condition (according to the length of time out of water and the probable approach of decomposition), temperature, and humidity. If too much salt is used, the fish will soon become hard and dry and their flavor will be acrid and unpleasant. If the quantity of salt is insufficient, the fish within a short time will become tainted and unfit for food. More salt must be used in warm weather than in cold; fat herring require more than lean, and small herring need less salt than large. As a general rule, from 60 to 80 pounds of salt are used to each barrel (200 pounds net weight) of herring. When the fish are thoroughly mixed with salt, they are also covered with saturated salt brine (100° salinometer) as this ensures more rapid curing.

The herring remain in the vat from 8 to 10 days or until they are struck, that is, until the salt has penetrated their bodies thoroughly. During this time, the fish in the vats are stirred with a wooden paddle top prevent them from sticking together and to ensure an even cure. More salt may be added when stirring, if the brine shows and appreciable decrease in strength. When the herring are cured and are ready for shipping, they are removed by dip net and piled on a packing table to allow them to drain. After several hours of draining they are then packed into barrels to a net weight of 200 pounds. The herring are usually packed on their backs, bellies up, and at a slight slant. The method of filling is much the same as in packing Scotch-cure herring, with each layer laid at right angles to the one preceding. The fish are packed on more of a slant than in the Scotch cure and not so tightly. A little salt is scattered between the fish and over each layer, about 25 pounds to the barrel. The top layer is packed with backs up and receives a little more salt than the others. The filled barrels are headed, a hole is bored in the bung of each, and as much 100° salinometer brine as possible is poured in through a funnel. The barrels are then ready for shipment. The loss of weight in curing round herring is small, ranging from 6 to 10 percent. An average of 211 pounds of fresh herring are required for a 200 pound barrel of cured fish (Jarvis 1943b).

SPLIT CURE

This method is followed in Newfoundland, Nova Scotia, and other coastal areas in eastern Canada, and in New England. It is of minor importance in New England but is an important method of salting herring in Newfoundland. Formerly, only the largest fish were given this cure, but recently, all sizes have been packed as split-cure herring.

They are first soaked in brine for a few minutes to set the scales and to make handling easier. The herring are then split down the belly to the vent. The head is usually left on and, if so, the gills are taken out. The fish are next cleaned and as a rule the milt or roe are removed with the viscera.

After cleaning, the herring are soaked in salt water or light brine for 2 or 3 hours to remove blood and slime. They are then drained for a few minutes and packed in butts or large barrels, backs down, with the belly cavities filled with salt. More salt is scattered over each layer, using from 30 to 40 pounds to each 100 pounds of fish. Split herring require less time to cure than round herring or about 1 week. When thoroughly struck, the fish are repacked in barrels, using the same general method described for round cure. The repacked barrels are headed up, filled with 100° salinometer brine, and are ready for shipment.

A few barrels of herring are sometimes prepared by a variation of the split-cure. A cut is made across each fish just back of the pectoral fins, removing the head. The cut is continued down the ventral side to the vent, removing most of the thin belly flesh. The kidney (the dark "blood streak" along the backbone) is scraped away, and the fish are thrown into the tank to soak. After an hour or two, the herring are removed and packed directly into barrels, backs down. Some salt is scattered over them as the fish are packed in the layer, and more salt is scattered between each layer. About 30 to 35 pounds of salt, mixed with 2 pounds of fine, white pepper are required to each 100 pounds of fish. After 4 to 6 days, the barrels are filled with additional layers of fish of the same day's cure as those already in the barrels. The barrels are then headed and filled with fresh filtered 100° salinometer brine.

Another variation of the split cure is occasionally prepared. Only large fat herring are used. The fish are washed in brine, then split down the back to open in one piece, mackerel style, taking care that the knife does not go all the way through the body. The gills, viscera and belly membranes are removed. The cleaned fish are soaked in 50° to 60° salinometer brine for 1 to 2 hours, then drained and packed in barrels, flesh side up, in the same style as mackerel. An average of 35 pounds of salt is used to each 100 pounds of fish. After about 4 days (when the fish have settled) additional layers of the same day's cure are added. The barrels are headed and filled with fresh 100° salinometer brine through a hole in the bung.

SCOTCH CURE

In Scotland, herring salting is that country's most important fishery industry. In normal times, Scotland exported more salt herring to the United States than was prepared in this country. The Scotch-cure method is also the principal style of cure employed in Sweden, Canada, and Alaska. A knowledge of the methods followed in the Scotch cure is, therefore, important to fish curers in the United States. The standard reference on the Scotch cure of herring is *The Art of Fish-curing*, by "Viking" (Duthie 1911), which should be consulted for a more detailed account of the method than is possible to give here.

Figure 20. – Scotch-cure herring showing arrangement of fish in barrel.

The Scotch-cure herring industry is located principally on the Shetland Islands, on the east coast of Scotland at Frazerburgh, Peterhead and Wick, and on the east Anglian coast at Yarmouth and Lowestoft. The herring are taken by "drifters," using a long chain of gill nets. They are taken from areas close inshore to a distance of 80 to 100 miles from shore. The herring are sold in the open market to the curers, using a basket measure known as a "cran."

When the fishermen empty their baskets into the curer's containers, the herring are mixed well and regularly "roused" with salt. The herring are then delivered to the curer's plant where they are held in bins or gibbing troughs until they are gibbed (gutted). This should always be done as soon as possible. The herring are gutted with a short-bladed knife, removing very little of the fish, only the pectoral fins and upward to the gills. The gutted herring are thrown into tubs or baskets, one for each size being packed, usually two to four. When sufficient herring accumulate in the tub they are "roused". This is considered the most important step in the Scotch-cure process. A few platefuls of salt are thrown over the gutted herring after which the packer stirs them around in the tub until every herring has come freely in contact with the salt.

When the herring have been thoroughly mixed with salt they are ready for packing. The herring are shaken to free them from surplus salt and a few handfuls are dropped gently into the bottom of a barrel, which is damp inside and has been previously inspected to make sure that it is tight. No salt should be spread in the bottom of the barrel. The method of filling as described in detail under the discussion of Alaska Scotch-cure herring, except that as a rule somewhat less salt is used in Scotland than in Alaska. Usually the herring are packed above the level of the barrel, as they sink rapidly in the salt. On the following, or second morning the fish will have sunk so that there is some space in the top of the barrel. This should be filled up level with the top, using herring of the same day's cure. The barrel is headed and laid on its side. After the barrels have laid on their sides for 8 or 10 days, a bunghole is bored in the bilge. The barrels are set on end, with the heads taken out and the pickle is allowed to run off through the bunghole.

The barrels are then filled with herring of the same day's cure. A barrel of herring is required to refill five barrels. Very little salt should be scattered between layers and none at all over the top layer. When the barrels have been properly filled, a little clear brine is thrown over the top layer to improve the appearance of the fish (fig. 20). After heading up, the barrels are completely filled with strong brine and the cure is considered complete.

There are two types of grades or brands in Scotland, the government of "Crown" brand and the private or "trade-mark" brands of curers who do not use the government brand. The reputation of Scotch-cure herring is built on the Crown brand, however, and the sizes as determined by the Scotch Fishery Board are representative of the industry. They are reported by Duthie (1911):

> **Lafull** Herrings intended for the crown lafull brand must be 'large, full fish of not less than 11 1/4 inches in extreme length' when cured. If on examination more than fifteen spent, torn, or broken, or more than fifteen undersized herrings are found in the original packing of a barrel, or more than six in the filling up, the officer is entitled to withhold the brand.
>
> **Full** For the 'full' brand the herrings should be 'full fish of not less than 10 1/4 inches in extreme length when cured. Eighteen objectionable herrings in the original packing and nine in the filling up constitute the limits allowed for this brand. For both the 'full' and 'lafull' brands the herrings should be quite full of milt and roe.
>
> **Matfull** 'Matfull' herrings are 9 1/4 inches long when cured, but in this case it is sufficient if milt or roe is clearly visible at neck or throat. If more than twenty-one defective herrings are found on inspection, or more than nine in the filling up, the officer is justified in rejecting the fish.
>
> **La Spent** For the 'La Spent' brand, the herrings should be not less than 10 1/4 inches in extreme length when cured. If more than eighteen objectionable herrings are found in the original packing, or more than nine in the filling up, the fish are not considered worthy of the brand.
>
> **Spent** The crown 'Spent' brand is applied to barrels of 'spent' or empty fish of less than 10 1/4 inches, but not less than 9 inches in extreme length, provided not more than 18 objectionable herrings are found in the original packing, or more than 9 in the filling up.
>
> **Mattie** The 'Mattie' brand is put upon barrels of small- sized herrings that are ineligible for any of the other brands. They must however, be at least 9 inches long, and the original packing should not contain more than 30 nor the filling up more than 12 torn or otherwise defective fish. To be eligible for this brand, however, fish must not have been caught on the coast of Shetland before the first of July, or on the east coast of Scotland before July 10; and the presence of an excess of oil in the barrels may also disqualify the herrings for the brand.

ALASKA SCOTCH-CURE METHOD

In the United States the "Scotch-Cure" method of salting herring is largely confined to Alaska, and differs in some details from the process as followed in the British Isles. The operational details, construction, sanitation, and other manufacturing factors also show differences. Therefore, it is felt that the Alaska industry merits a detailed discussion, especially of the points that are not the same as the industry in Scotland.

Condition

Herring intended for Scotch-cure must be strictly fresh and in good condition. They must be as free as possible from "feed" or other materials, causing enzymic spoilage. The herring should not be bruised, crushed, or piled deeply in the hold of the fishing boat. Care should be taken to drain off as much sea water as possible in brailing. If the fish soak in sea water for several hours in the hold of the fishing boat they do not make good Scotch-cure herring. A large quantity of Alaska herring is unfitted for curing because too much sea water has been included with the fish. No ice should be used on herring for this cure and they should be processed immediately on landing. Thin or small-sized fish should not be used.

Gibbing

The knife used in gibbing herring is much like a paring knife, but with a short blade, 2 inches long. It has a four inch handle. In Alaska, girls wear canvas gloves while gibbing to guard against cutting the left thumb, and for greater ease in handling fish. Scottish girls merely wrap up the left thumb and work with bare hands. When one sees how slimy and dirty canvas gloves become, the Scottish method seems better.

The use of automatic sorters when unloading is common. In most salteries the gibbers are no longer required to spend considerable time sorting out the herring too small for salting. The fish are not washed before salting, for washing spoils herring for Scotch-cure.

In gibbing, the herring is taken by the middle with the left hand, thumb on one side of the head, fingers on the other, leaving the throat clear. The knife is stuck through the gills, just under the gill cover. Then, with the edge of the blade toward the gibber, a sharp twist is given upward and outward. If properly done on fresh herring, the throat and pectoral fins together with the main gut, heart, liver, and gills, will be removed at one time. At first, it requires more than one move to gib herring in this way, but with practice the knack is soon acquired. In drawing out the intestine, it is better to pull outward more than upward; otherwise the herring are apt to be torn, especially if they are fat.

Rousing

When the fish are gibbed, they are thrown into rectangular rousing bins, which are placed back of the gibbers. As they are gibbed, all herring are sorted into sizes that are being packed at one time. These may vary from two to four different grades. Each gibber usually has two bins and when more than two sizes are packed, partitions are placed in the bins. In Scotland, the gibbers work in crews of three but in Alaska they usually work singly, though they may sometimes work in pairs.

Rousing the herring is one of the important points of the Scotch-cure method. In the rousing bins, the fish are mixed thoroughly with a specially manufactured kiln-dried rousing salt. A very coarse salt must never be used for rousing. Plenty of salt should be thrown over the herring, to be worked thoroughly and carefully through them by hand. Every part of the herring should come into contact with the salt as completely as possible. Double-washed California salt is generally used in Alaska. The herring are packed into the barrel immediately after rousing.

Grading

Scotch-cure herring are graded into the following sizes in Alaska:

Cured Size	Full fish	Empty Fat fish (Matjes)	No of fish in 250 lb. barrel
12 1/4+	Extra large full	Extra large	400 - 500
11 1/4+	Large full	Large	600 - 650
10 1/4+	Full	Medium matties	700 - 800
9 1/4 +	Matful	Select	850 - 1000

Herring shrink an average of one-quarter inch in curing; therefore, an allowance of 1/4 inch must be made when grading fresh herring for length before curing (Jarvis 1936a).

If the herring have not been properly roused they will be in poor condition when the barrels are opened for inspection. The fish will be sticking together in the layer and when separated they will be scaleless and discolored. Some "rusting" or oxidation indicated by a brown gum or deposit adhering to the skin will have occurred. In this connection it is important that the herring be packed as soon as possible after rousing. If this is not done, some of the original pickle will form in the rousing bin and will be lost. Fish not properly roused are usually too hard-cured and do not have the desired flavor or texture.

The standard Scotch-cure herring barrel weighs 250 pounds, net weight of herring, is made of spruce staves 3/4 inch thick, is 30 inches high, has a 17 inch head, and a capacity of 32 U.S. gallons (26 2/3 Imperial gal.). The largest portion of the Scotch-cure herring packed in Alaska, however, is put up in half-barrels holding 125 pounds, net weight, of fish. All barrels should be new, of best quality, should be inspected carefully by the cooper for leaks or any other possible defects, and soaked before use.

Packing

Herring intended for Scotch cure are never washed before being gutted and salted, either with water or brine, and are never repacked. They are cured in the original barrel and in their own blood pickle. No salt is sprinkled in the bottom of the barrel when packing or "laying" herring. If the fish have been properly roused, enough salt will cling to the bodies. Care should be taken when lifting the herring from the rousing bin not to carry any of the loose or surplus salt into the barrel as this makes for a non-uniform salting and packing. Some curers shake the fish gently when lifting them from the rousing bin to get rid of surplus salt.

The packer begins by taking one roused herring and placing it, back down, against the side of the barrel, straight up, not at a slant. Two other herring are placed against the first, their heads to the side of the barrel and their tails meeting, or overlapping. A middle herring is placed in front of the tails of the last two, then, two others, one to the right, the other to the left, with heads pointing against the barrel staves, and so on, until the layer is completed. The position of the middle herring is alternated so that the head of the first points to the right while the second middle herring faces the left (fig. 20). Particular attention must be paid to keeping the herring lined straight up, packed tightly, and the rows regular. If the herring are packed on a slant, with the rows irregular, the curing will be uneven, as one side of the herring will get more salt than the other. The space at the side of the barrel caused by the meeting of the heads of the herring must be filled by placing two herring over the heads of those making the hollow. These two fish, known as head herring, are placed so that their heads point in opposite directions. This leaves an even surface for each of the succeeding tiers.

Salt is then sprinkled over the entire layer. The amount depends on the season of the year, the market for which the fish are cured, and the quality and size of the herring. Duthie (1911) states that as a general rule 1 barrel of salt is required to cure 3 barrels of herring.

> "A safe rule to follow is to never entirely cover the bellies of the fish of each tier, but give them enough salt to almost cover the bellies, with the exception of extra large herring, especially those with milt and roe which require more salt. Scotch-cured herring when *thoroughly cured* (say after the 14th day) should be free of undissolved salt. Remember that large herring require more salt than small ones, fat herring more than those of poorer quality, milt or roe fish more than empty fat ones of *matje quality*. (Slinks or spents do not come under the matje selection.) Milt herring require more salt than roe fish and remain softer than spawn herring. When the first tier has just been packed and salted, continue to pack the second tier *by crossing the first tier at right angles*, keeping the fish belly up. Prevent the lines from curving as this spoils the appearance of the pack. Salt as before and continue packing, salting and crossing each tier until the barrel is full. A little more salt may be added to the tier in the center of the barrel as this helps to keep up the strength of the pickle (Klie 1920)".

Curing

Some packers allow the brine to be formed solely from the moisture extracted from the fish with the added salt, while others add brine. Klie (1920) states that brine should be added in warm weather as it strikes in from both the outside and inside of the fish more quickly than if the brine were left to be formed from the fish, thus reducing any possibility of spoilage. The barrels are usually headed immediately after packing, laid on their sides, and filled with 100° salinometer (saturated) brine through a hole bored in the bilge.

The barrels should always be stored on their sides as the herring cure better in this position, because it allows the salt and brine to strike through the fish more evenly and freely. For some reason, all barrels should be rolled half-way around occasionally. At this stage, the herring are said to be "in sea stick." From 10 to 14 days are required to cure the herring thoroughly but the barrels are usually refilled sooner.

Refilling

When the barrels are to be refilled they are set on end, the heads are taken off, and the brine or pickle is allowed to run out through the bunghole. The brine is sometimes saved for rouse. The herring will have sunk considerably in the barrel, which must now be filled up. One barrel of herring to every five is required in refilling. The herring used for refilling should be of the same lot as the barrels to be refilled and they should be well washed in fresh, clean brine.

The herring are filled in as described for the original packing but must be handled more slowly and carefully. The usual way is to straighten out each herring between the thumb and forefinger. This not only straightens the fish but also flattens out the belly. Very little salt is added between each layer and none is used on top.

The top layer must present a very attractive appearance as the pack is judged a good deal by the condition of this layer. The bellies of the herring in the top layer must be well flattened out and the fish set straight up on their backs. To get a full barrel, the herring must show above the croze. When the top layer is completed, the heads are pressed down on each side of the barrel, and three head herring are laid in on each side to make the layer even. These must also be packed straight up with backs down. The head of the barrel is then "jumped in." This usually requires considerable pressure but flattens out the herring on top and leaves no space which would allow the fish to shake about in transportation. This practice is not followed in packing summer herring. The barrel is laid on its side to be filled with brine testing from 80° to 100° salinometer. This may be the original pickle, which has been strained or fresh pickle may be used. The bung is then driven in and the cure is complete.

Branding

When the head of the container is dry, it is stenciled with the name of the packer, trade mark or brand name, grade, and place of cure. According to law the weight of the herring must be marked on each container. This is usually stamped on the bottom end, 250 pounds net, for barrels, and 125 pounds net for half barrels. A few packers also mark the gross and tare. This practice is not followed regularly.

Transportation and Storage

The cured herring are moved out of the saltery warehouse as soon as possible and shipped to Seattle, the primary marketing center, by steamship. In shipping, the barrels must be stowed carefully and must be kept away from the boilers or steam pipes. On landing, the herring are unloaded into warehouses, where they are held in chill storage at 34° to 36° F. until sold. They must be kept in chill storage, as they are mild-cured; depending, in part, for flavor on the "blood pickle," which ferments and sours easily at higher temperatures.

DUTCH-CURE HERRING

Herring are usually caught in drift nets (gill nets) a considerable distance out to sea in the Netherlands. Whenever possible the gibbing is done as the nets are hauled on board and the herring are picked out. The Netherlands way of gibbing and cutting is very little different from the Scotch, except that no attempt is made to pull out the crown gut, as it is believed that too much of the fat which gives the fish a nice flavor is removed at the same time. The opening at the throat is also a little smaller that for the Scotch cure herring. As the herring are dressed they are sorted into baskets according to size. When there are about 200 herring in a basket it is dumped into a large tray or shallow box filled with fine Lisbon salt. The gutted fish are mixed or roused thoroughly with the salt, which, in this instance, is called "dredging" or "rolling."

When the herring have been rolled sufficiently in salt they are packed in barrels, backs down, using the same method as for Scotch-cure herring. A little Lisbon salt is thrown between each layer. The amount of salt used, as a rule, is slightly less than for Scotch-cure herring. When the barrels have been filled, a bucket of "blood pickle" is poured into each barrel. This pickle is a brine made from the gibs and blood removed in dressing. These are saved and mixed together with sea water and salt to make the pickle. After adding the pickle the barrels are headed up at once and stored in the hold of the fishing vessel.

The barrels are headed at once so that the herring will be affected by air as little as possible. The barrels are left in the hold for from 6 to 10 days, when the herring should be struck through. The barrels are taken on deck again, the brine is drawn off, and the barrels are filled with herring of the same day's pack. The original brine is strained and put back in the barrels. One barrel in four is lost in this filling or sea-packing, as the Netherlanders pack compactly. In this cure it is considered important not to do this filling too early, lest the herring be shriveled, but if delayed too long, the herring will be battered about, injuring their appearance.

When the fishing boats return to port, the heads are taken off the barrels, and the brine is drawn off. The barrels are refilled with herring in final preparation for shipment. In this last filling it is estimated that 14 barrels of herring as landed will give 13 barrels ready for shipment. The filled barrels are headed and "tightened," after which the original pickle is poured in through the bunghole. At this time a great deal of the herring is selected or graded and repacked in kegs of various sizes. A small keg, one-sixteenth of a barrel, holding from 40 to 50 herring, is a favorite size, especially for the American market. Another small keg holding about 12 herring is packed for the American trade. This last size is known in the American delicatessen trade as "milker" herring. A Dutch barrel of full selected herring with milt and roe holds about 800 fish, and weighs from 110 to 115 kilograms net (242.2 to 254.6 lbs.), exclusive of salt and brine. Some selected herring are also packed in Scotch barrels

Dutch-cure herring are graded first, according to the degree of spawning development, into four grades: Full herring (vol herring, with barrels branded "VOL"; matties (maatjis, branded "M"); spent herring (ijlen, branded "IJ"; and herring which have recently spawned (ruit, branded "KZ"). Each of these grades is further divided for quality into three subgrades. The barrels are marked 1,2, or 3 according to grade. For example, a barrel of first-class full herring would be marked "VOL 1." Herring cured the same day when caught, in good condition and free from faults, are graded as 1 or 2 according to quality and treatment. Herring which could not be packed the same day when caught (because of a large catch) but were cured the next day, together with chafed fish, or herring with torn bellies, are graded as number 3. Sometimes barrels holding herring not packed the same day when caught are marked "O". Dutch-cure herring packed in Scotch barrels are graded as follows: Superior, 600 to 650 to a barrel; Sortierte, 700 to 750; Prima, 800 to 850; and Small full, 900 to 1,000. The Netherlanders also distinguish between the quality of fish caught out at sea or near the coast, and the barrels are generally marked to show this.

NORWEGIAN-CURE HERRING

In Norway the greater proportion of the herring catch is made in the fjords by a large seine known as a "landvade." When a school is found, one end of the net is fastened on shore while the remainder is used to surround the fish. The seine is gradually drawn in until the fish can be pocketed behind the net in a cove or under the lee of a headland. The herring usually are held in the pocket at least 24 hours because they are often full of feed when caught. After 24 hours the digestive tract is empty and the herring are not so susceptable to enzymic deterioration on the way to the saltery and during the first stages of the curing process. The fish are usually cured in the vicinity of the place where they are caught, so they can be put in cure when almost alive. This gives the Norwegian cure herring an advantage in quality over the Scotch- and Dutch-cure herring. The herring may be held in the nets for weeks if necessary, removing at one time only enough for the daily curing requirements.

When the herring are brought ashore the first step in preparation is gibbing. As seine-caught herring are mixed in size each gibber has from three to five tubs, one for each size being packed, into which the herring are sorted as they are gibbed. Although the herring can be gibbed for the Norwegian cure with the fingers or a small knife, as a rule a specially developed scissors is used. Only a small cut is made at the throat, much smaller than for the Scotch cure. The fish are bled rather than gutted. A small triangular piece including the pectoral fins is removed, just large enough to include the heart. The cut must be deep enough to cut into the main blood vessels, close to the neckbone, in order to remove the blood they contain. The gills are not removed except from full or spawning herring. The sizes into which the herring are sorted, with the grade marks, are as follows:

Grade mark	Size of cured fish (inches)	No. of fish in 100 kg. barrel
KKKK	12 1/4 +	425 - 475
KKK	11 1/4 +	550 - 600
KK	10 1/2 +	700 - 800
K	9 1/2 +	800 - 900
M	9 1/2 +	more then 900

No salt is thrown in the tubs with the gibbed herring. After sufficient fish of any one size have been gibbed to make it worth while to handle, a salter packs them in new barrels that have been well soaked in sea water. A scattering of salt is thrown on the bottom. The herring are then laid in, packed on their backs, slantwise, one fish lying across the other throughout the layer. This is in contrast to the Scotch method where the fish are packed straight up, on their backs. The Norwegian-cure fish are also packed much more loosely than the Scotch.

The method of arranging the layers in the container is otherwise much the same. As the herring are packed in, salt is scattered between them in layers, and a thin layer of salt is scattered between layers of fish, using a total of about one-third as much salt as fish. The uppermost layers are often packed in, backs up.

Some packers add two gallons of saturated salt brine to each container as soon as it is filled and head the barrel immediately. Others do not add brine or head the barrels until one day after packing. Before the barrels are headed, a layer or two of herring are generally put into them, in order to fill the space created by the shrinkage of the herring. If the brine is added immediately after packing it is believed that the salt dissolves more quickly, hastening the rate of penetration through the flesh. The barrels are allowed to stand from 10 to 14 days after heading. At the end of this time they should be completely cured and ready for repacking.

In repacking, the first step is to remove each barrel head and drain off some of the "pickle", which is saved. The fish are not taken out of the barrels in repacking. This consists of adding additional layers, backs up, in the top, filling the space created by the shrinkage of the fish in curing and by draining off the brine. The method of fill is the same as in the original salting. Fish of the same day's cure is used for repacking and a little salt is scattered between each layer of fish. In repacking, 4 barrels of the original cure are required for 3 1/4 to 3 1/2 barrels of the finished cure, or a loss of 18.8 to 22.5 percent. The filled, repacked barrels are headed immediately and filled with strong brine. The original brine or pickle drained off for refilling is usually strained and used again. If there is not enough of the original brine, fresh brine is made and added. The Norwegian barrel is one Imperial gallon smaller than the Scotch barrel in capacity. While the Norwegian-cure herring is more heavily salted than the Scotch-cure, the difference is not so marked as it was. Also in contrast to the Scotch-cure, these herring contain more fat than milt or roe.

MATJE CURE HERRING

The matje cure is a very mild cure applied to empty fat fish, that is, fish not in a spawning condition. The stomachs are usually full of feed, composed of small shrimp-like organisms. Some katje cure herring have been packed in Alaska, but the complaint is made in New York and other Atlantic coast areas that this pack is too heavily cured. The best matje-cure fish come from the Hebrides, followed by those from the Shetland Isles, off the coast of Scotland. The following description outlines Scottish practice in preparing the matje cure. The herring require careful curing, begun as soon as possible after catching. The fish should not be left exposed to the sun while awaiting cure and they must be handled as little and as carefully as possible. If the herring are weighted down or handled at all roughly, they become bruised and softened, making them unfit for the cure. Gutting must be done very carefully and thoroughly, otherwise the stomach contents and fat adhering to the crown gut may cause rapid deterioration.

As a rule, two sizes of herring are selected, large and medium. Small, torn, or broken fish are discarded. English second fishery-salt (equivalent to the three-quarter-ground salt of the Pacific coast) is used for rousing, and Spanish salt is scattered between the layers. The roused fish are shaken to rid them of surplus salt and the amount of salt added to the barrels is about half the usual quantity for the Scotch-cure. The rousing tubs must be emptied frequently to get rid of the used salt and keep the herring clean. Rousing tubs should have several holes bored in the bottom to allow the blood to escape.

Immediately after a barrel is filled, a fresh, clear, "made" brine testing about 90° salinometer, should be poured into the top, about a water bucket to a full barrel, half that quantity to a half-barrel. The barrels should then be stood on end in a shady curing-shed. After standing for two nights the containers are refilled. The surplus pickle is usually poured off over the rim of the barrel; not through the bunghole. The barrels are not filled quite so tightly as for the ordinary Scotch cure, and while there is usually a certain tolerance for fish that do not meet the grade, in the matje cure all herring that are in the least bit objectionable should be discarded. When the barrels have been refilled they are headed immediately, filled with brine through the bunghole and stored in a cool, dry shed. As far as the United States is concerned, chill storage is essential to handling matje cure successfully. As soon as the herring barrels are refilled they should be put in storage at 34° to 36° F. The fish should be held continuously in refrigeration until shipped to the retail market.

Chapter Fourteen

SALMON SALTING

MILD CURE

Mild-cured salmon is a lightly salted product which is largely dependent on refrigeration for preservation. The method of curing was first introduced on the Pacific coast in 1889 (Cobb 1930) when a shipment was prepared for the German market, but the experiment was unsuccessful. Salmon was not mild-cured in any quantity until 1898, when two small plants were established on the Columbia River. Packing of mild-cured salmon began on Puget Sound in 1901. While a few tierces were occasionally packed in Alaska prior to 1906, it was not until that year that mild- curing was established on a commercial basis. A large part of the king salmon taken in southeastern Alaska is now mild- cured.

Mild-cured salmon must be considered an intermediate of half-finished product since a large proportion of the amount cured is used to make smoked salmon. Some of the pack was formerly sent to Germany and the Scandinavian countries for this purpose. Two world wars have disrupted foreign trade but meanwhile markets in New York, Philadelphia, Chicago, and other large cities, have absorbed a considerable quantity of the pack.

Mild-cured salmon must be handled more carefully than any other salmon product. In few other food products is handling so important in determining the quality of the finished article. Red-fleshed king salmon (*Oncorhynchus tschawytscha*) is used almost exclusively and dressed fish weighing 18 to 20 pounds are usually regarded as the smallest sizes suitable for mild-curing. There is some variation in this minimum, as at Astoria, Oregon, fish of less than 30 pounds in weight are usually rejected by mild-curers, while in Vancouver, British Columbia, Canada, the minimum size is 18 pounds, dressed weight. From time to time, packs of mild-cured chum and pink salmon have been put up, but have not found a market. Coho or silver salmon (*O. kisutch*), is the only other species utilized to any extent for the manufacture of mild-cured salmon and is usually prepared to fill market demands for a lower-priced smoked salmon. In 1944, according to the Pacific Fisherman, 1,050 tierces of silver salmon were packed in North America. Tierces average 825 pounds net weight of fish, exclusive of brine.

Salmon intended for mild-curing must meet certain requirements as to quality: Fish must be (1) strictly fresh; (2) reasonably fat; and in good condition (thin fish are not wanted); (3) the skin must be bright, there must be no "watermarks" or other blemishes; (4) the flesh must not be pale in color; (5) the flesh must not be bruised or broken and there must be no pew marks or other signs of rough handling; and (6) they must not be belly-burnt, that is, show signs of softening in the abdominal region (Jarvis 1936b).

For this reason troll-caught salmon intended for mild-curing are always gutted when caught and packed in crushed ice aboard ship. Salmon taken by other gear are often gutted, or at least, packed three or four together in a box filled with crushed ice, which acts as a refrigerant, and aids in drawing out the blood. The boxes may be piled up in several tiers in the hold of the boat but the weight is distributed, and individual fish are not weighted down more heavily than by the other fish in the box.

Butchering

The first step in preparing mild-cured salmon is known as butchering. The butcher removes the head, cutting from the back, leaving as much as possible of the bony structure just above and below the gills. This makes the fish stand up better under handling. If the bony structure were cut away, the sides would break easily in curing, and the hooked on which the fish are hung during smoking would be more likely to tear out.

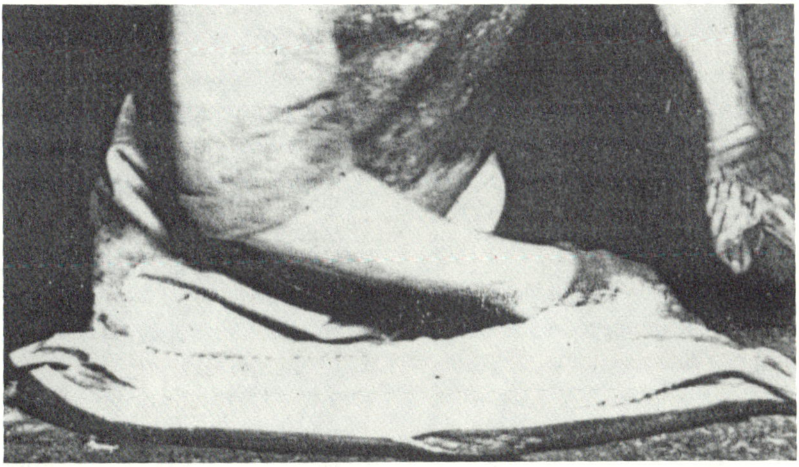

Figure 21.—Splitting salmon for mild cure.

The fish is then scored with three or four cuts along the lateral line. These are made just through the skin, but should not penetrate into the red meat. Scoring allows the salt to penetrate more rapidly, insuring a better cure. A specially designed star-pointed wheel is sometimes used for this purpose. It makes a series of small cuts varying from half an inch in length at the tail, to one and one-half inches at the shoulder. A number of extra cuts or scores are made if the salmon is large. After scoring, if the fish is not already gutted, it is split down the belly to the vent. The viscera are removed, as are most of the belly membranes, and a cut is made along either side of the kidney, the dark red mass found just below the backbone, at the top of the belly cavity. The kidney is then scraped away as completely as possible.

After cleaning, the salmon is ready for the splitter, who is the most important workman in a mild-curing establishment. The grading of mild-cured salmon and the percentage of loss of raw material, depend largely on the skill of the splitter. An unskilled or careless workman is often responsible for considerable losses. A specially shaped knife is sometimes used in splitting, the end of the blade being nearly square, but the type of knife used depends on the preference of the splitter. he splitter turns the fish on its side, nape to his right, and with the open belly toward him, and then forces the shoulder down on a sharp-pointed nail so that the fish will not slip. He first makes a cross-cut at the root of the tail, but no farther. Short incisions are then made below the anal fin, cutting just above and below the backbone. Then, with the upper “lug” or shoulder tip of the fish in his left hand, the splitter enters his knife at the shoulder above the backbone, and holding the blade steady with the edge at a slight downward angle touching the bone, takes the whole side off with one sweep of the knife (fig. 21). This can be done better if the fins are taken off before gutting. In some plants the fins are removed. If the work has been well done, little flesh will be left on the backbone and the sides will be smooth. A thin line of bone will show down the center of the side, which increases the value of the finished product.

To cut the second half loose from the backbone, a cut is made at the shoulder just under then bone. With one sweep of the knife, the edge of the blade resting against the bone at a slight upward angle, the splitter separates the backbone from the flesh down to the root of the

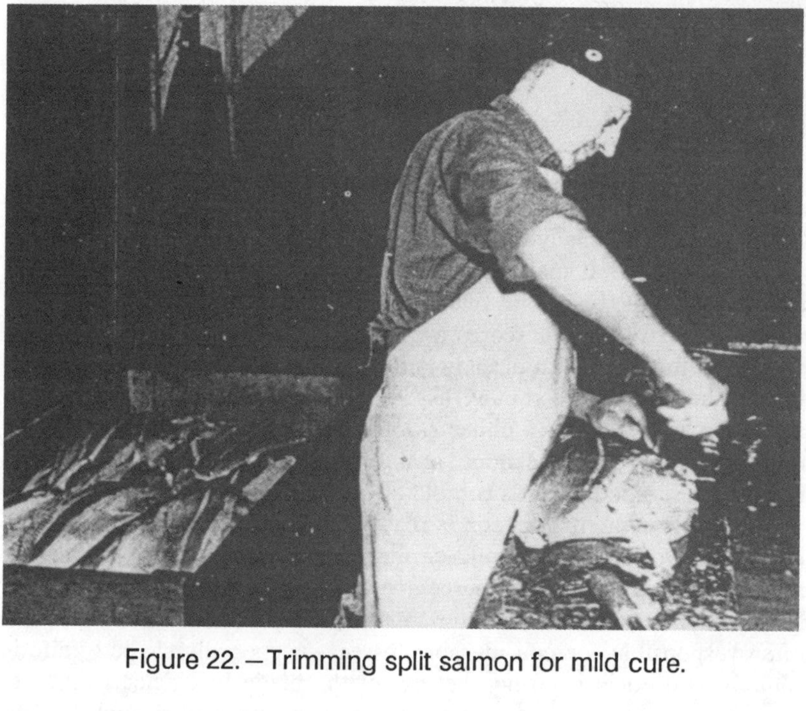

Figure 22.—Trimming split salmon for mild cure.

tail, without removing the fish from the nail. As with the first half, little flesh should be left adhering to the bone and a film of bone should show down the center. In other words, the two sides should be exactly alike.

Figure 23.—Salting salmon for mild cure.

Washing, Trimming, and Chilling

The sides are washed thoroughly in cold water and then passed onto the sliming table where they are laid skin side down with the thin, or belly edge, toward the front. All blood clots, loose membranes, and fragments of bone are removed. Any blood remaining in the veins along the abdominal cavity is scraped off by pressing it toward the back of the fish either with the fingers or the back of the knife blade.

If the blood is not squeezed out in this way the salt will harden it during the process of curing, causing discoloration of the flesh and lowering of the value of the salmon. Any slight necessary trimming, may also be done at this time (fig. 22). Great care must be taken in handling the newly split sides, as they are very tender and may be easily broken or bruised. In lifting them by the lug or collarbone, the curer should have his fingers to the inside and his thumb to the skin side, as otherwise the flesh may be broken.

Figure 24.—Packing sides of mild-cure salmon in tierce after salting.

From the slimers the sides are taken to a tank of ice-water or iced brine. Warm water tends to loosen up the muscle flakes if the salmon is left too long. This tank is known as the chilling, or more commonly, the sliming tank. The latter name is a misnomer as all slime should be removed before the sides go into this tank. The object of this step is to prepare the sides for curing and it may be likened to case hardening. Unchilled sides would absorb too much brine and the penetration would be too rapid during the first part of the cure. Chilling also serves to draw out the blood , improving the color, and also helps to prevent oil from oozing out of the flesh, which is apt to occur where an amount of cut surface is exposed, especially under pressure during curing. There is some variation in the length of time the sides are left in the sliming tank. In some localities, the period is two hours; in others from one-half hour to one hour. The temperature of the brine may vary from 30° to 40° F. and its salinity from 60° to 70° salinometer. The brine should be made with fresh drinking water, boiled and strained before use, and should be changed at least daily.

After sliming, the sides are drained. This is done in another tank, or the salmon may be placed on a two-wheel cart or portable table. The fish are transported more easily and it is claimed that the water drains off better where the second method is used. On the Columbia River, a cart holds just a tierceful of sides, 7 lengthwise of the cart, and 3 at the end, or 10 sides to a layer. This arrangement helps in counting the number of sides going into a tierce.

Salting and Packing

When the salmon sides have been drained sufficiently, they are taken to the salter, who works from a special bin or box of convenient height filled with fine salt of the "dairy" type. A special grade of salt known as "mild-cure" is usually demanded, which meets the requirements of a low content of chemicals other than sodium chloride, contains no organic impurities, and is of small, even grain. The salmon is taken one piece at a time and placed in a salt box, skin side down. Salt is scooped over the side with the hands but it must not be rubbed or pressed into the flesh of the fish, as this might injure the texture and cause absorption of more salt than necessary (fig. 23). Sufficient salt always adheres to the surface by this method. The sides are picked up by the tips, and excess salt is allowed to fall back into the box. They are then packed in a container known as a tierce (Scofield 1925).

A tierce is a large barrel, made from fir or spruce and bound by 6 galvanized iron hoops. It holds between 800 and 900 pounds of fish, with an average of around 825 pounds, cured weight. The gross weight, including pickle, ranges between 1,100 and 1,200 pounds. A few handfuls of salt are thrown on the bottom of the tierce, followed by a layer of salmon sides, skin side down. In packing, two sides of fish are laid close to opposite sides of the tierce, alternating head and tail, the back or thick part of each side being placed against the side of the tierce. Other sides of salmon are packed from the sides of the tierce toward the center, napes and tails alternately, the back of each side being drawn half-way up and resting on the side already laid. When complete, the layer should show a level surface, this depending a good deal on how the last or center piece is laid fig. 24). A little salt is scattered over each layer, which should be laid at right angles to the one preceding. The top layer should be packed skin side up, and a little more salt should be scattered on this layer than on the others. The amount of salt used varies from 85 to 120 pounds to the tierce. One of the leading Canadian mild-curers uses 90 pounds of salt to the tierce, and this may be taken as the average. Some curers use as much as 15 pounds of salt per 100 pounds of fish.

There is some variation in the curing process at this stage. The tierce is filled only to the croze and in some districts it is headed at once and filed with a 90° to 95° salinometer brine until the tierce will hold no more. In others, the tierce is held from 24 to 48 hours before heading, after which it is headed and filled with 100° salinometer brine. The pickle or brine should be made from the same salt used for rousing and packing the fish. The water used in making the brine should be clear, pure, in fact, drinking water. Before using, the pickle should be strained through a fine sieve or a piece of clean cheese cloth, to free it

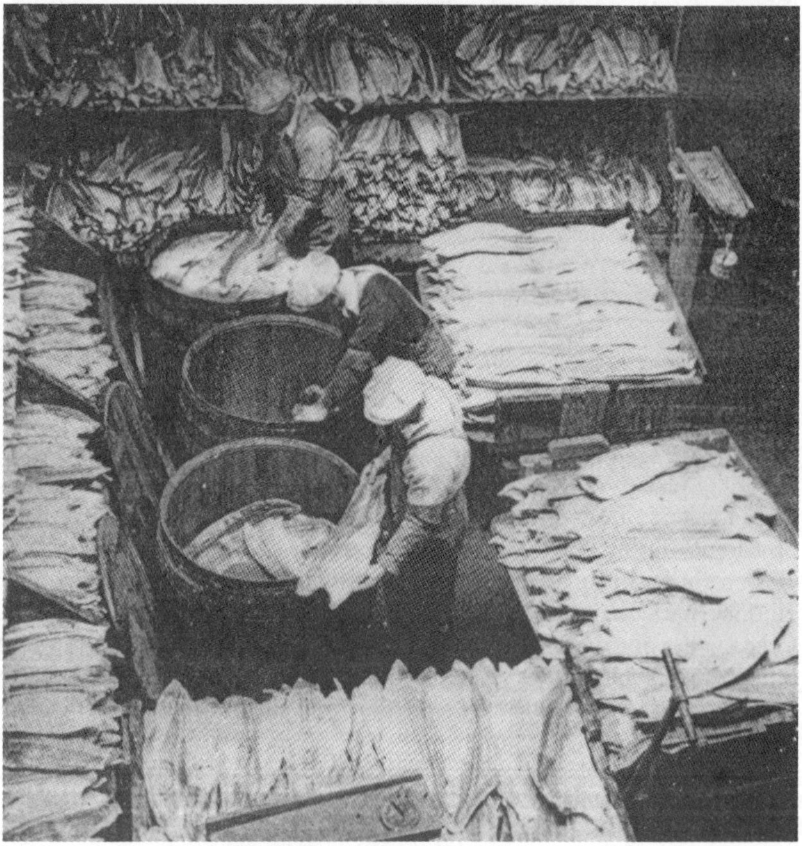

Figure 25. – Grading and repacking mild-cure salmon.

of any froth, dirt, or sediment. The strength of the brine is then determined by a salinometer. A centigrade scale salinometer is used by most mild-curers. The brine is usually made up to a strength of 90° salinometer, but during the first week or 10 days of the cure while moisture is being extracted, it decreases in strength to about 70° salinometer. After repacking, the strength of the brine should not fall below 85° salinometer, and it should hold this strength for some time.

After the tierces have been headed and filled with pickle, they are usually rolled into a room where the temperature can be kept down to 32° to 34° F. Here they are left to cure, stored in rows one or two tierces high. The temperature of the storage room should be held constant and not allowed to fluctuate, as this causes oil to exude from the flesh, escaping into the brine. The tierces are not always rolled into the chill room immediately after packing. Some curers, especially those working in cooler climates, leave the tierces in the packing room for 4 days, then send them to the chill room for 10 to 20 days before repacking.

If the tierces are not kept full of pickle, the sides of fish may be shaken about and broken when the tierces are shifted while being inspected, at intervals, to determine whether there are any leaks in the containers. No tierce is perfectly tight at first, and the staves absorb some brine. If any part of the fish is left uncovered by the brine, yellow, discolored spots develop, so-called "rust spots", which lower the quality of the finished product. Therefore, it is extremely important to see that the tierces are kept full of brine during the curing period and also after repacking. A tierce of salmon may absorb several gallons of pickle in the first 2 or 3 weeks of cure, especially if the fish are "dry."

The amount of shrinkage during the first 3 weeks, before repacking, is estimated to be about 30 percent. Less shrinkage occurs in fat ocean-caught fish. Thin "dry" fish, especially those caught when well on their way to the spawning ground, may shrink as much as 35 percent in weight.

Repacking

After the salmon have been held in storage at least 20 but not more than 90 days, they are repacked. The tierces are rolled out and unheaded. Each piece is taken out carefully, holding the sides with the fingers on the flesh side and the thumb on the outer or skin side. The sides are sponged or cleaned carefully, removing all salt or other extraneous material on the surface. Either ice water or chilled brine is used to wash the sides of salmon, depending on condition. If the fish are soft and rather poor (thin), they should be washed in brine, but if the sides are firm and thick, ice water may be used. It is the opinion of some curers that chilled brine should always be used.

The next step is weighing and grading the sides. Unlike curers in other districts, those on the Columbia River grade twice. While the fish are being dressed three chilling tanks are used, one for each size. A rough grading into large, medium, and small sides is thus obtained. This is an advantage in packing and curing, as the time required for curing varies with the size of the side; much work is also saved in sorting for repacking. When repacking, a careful separation is made into from 6 to 10 grades (fig. 25). The designations of the grades depend on the number of sides needed to fill a tierce, and are expressed as 40 to 50, 50 to 60, 60 to 80, 80 to 100, and 100 to 120 sides per tierce. Slightly broken sides are graded as "B" of that size, while other more defective sides are placed in a third grade and called culls. Color of the sides is also considered in grading, and pale or off-color sides are segregated. The system of grading differs somewhat in various districts but this description indicates the general method.

In repacking, the sides of fish should be replaced as nearly as possible in their original position, those curved in shape being placed against the sides of the container, and straight pieces laid in the middle of the layer. No salt is used in repacking but as soon as the tierce is filled, the head put in, and an examination made to determine the tightness of the container, it is laid on the side opposite the bung, and filled with ice-cold pickle made to a strength of 90° to 95° salinometer. The tierce will contain about 825 pounds of salmon after repacking, and some 14 gallons (U.S.) of brine may be required to fill it. The gross weight will average 1,100 pounds. The tierce is then put back into chill storage and filled up daily for a week or more with pickle through the bunghole. If mild-cured salmon is stored for any length of time, the tierces must be tested at frequent intervals for leakage.

The head of each tierce is marked to show the number of the tierce (consecutively); the number of salmon sides in the tierce; the net weight, and the initial or brand of the packer. In some districts the tierce is marked with the packer's brand or initials, place where packed, number of sides of salmon in tierce, tare, gross and net weights, quality of fish (I, II, III or T), and size of fish L (large), M (medium), or S (small). In Vancouver, British Columbia, thin or broken sides are designated by the letter "X". If the salmon is of first quality no special mark is necessary, but second and third quality fish are always designated.

Storage and Use of Preservatives

Mild-cure salmon must be shipped under refrigeration and held in cold storage at all times. It is held at a temperature of 32° to 34° F., after repacking. If the salmon is to be kept more than three months some packers hold it at a temperature of 28° F. The temperature of storage must not be low enough to freeze the fish. The tierces must be inspected at least once a month, preferably every other week, while in storage. The smoker should leave the mild-cure salmon in storage until the day it is required for smoking. If the smoker must hold some stocks of mild-cure salmon in the smoking plant for longer than a few days, he should be supplied with storage facilities not more than 40° F. The salt cure is not of sufficient strength to delay spoilage for more than a brief period.

In the earliest days of the mild-cure salmon industry attempts were made to reduce or eliminate refrigeration by adding various preservatives when curing the fish. These were usually preparations of salicylic or boric acid, and the benzoates or other chemical compounds such as Urotropin, now limited in use or prohibited by food and drug administrations, here and abroad. The use of these agents was soon abandoned, in most cases before they were legally controlled, as it was found that the quality was adversely affected, and was causing mild-cured salmon to become unfavorably regarded by salmon smokers.

SALMON, PICKLED OR HARD-SALTED

Production, Species Salted, Gear Used

Pickling or brine-salting was the earliest commercial method for the utilization of salmon. The trade first became important in the early 18th century. Under conditions of transportation and methods of food handling of that time only a salted product could reach London from the Scottish and Irish salmon fisheries. At almost the same time the salmon of the Atlantic coast of North America were brine-salted, some for export to England, the rest for local sale. At one time, salmon were salted as far south as the Hudson River. A small amount of salmon (*Salmo salar*) is still salted on the Atlantic coast of North America in Nova Scotia, Newfoundland, Labrador, and Hudson Bay. The trade is small now, averaging a total of 3,000 barrels per year before World War II. Accurate figures for the last few years are not available, but it is believed that only a fraction of that amount is now cured. The Atlantic salt salmon is sometimes known as "Halifax salmon" on the New

York market. It usually sells several dollars a barrel higher than Pacific coast hard-salted salmon and is considered a fancier article. The method of salting is much the same as that used for curing hard-salted salmon on the Pacific coast.

Long before any permanent settlement was made, our ships visited the northern Pacific coast to put up cargoes of salted salmon which were later traded in Hawaii for sandalwood, which in turn was sold in China for furs, silk, teas, spices, or other Oriental goods (Cobb 1930). Today Hawaii remains one of the principal markets for hard-salt salmon. Pickled or hard-salted salmon is packed on a commercial scale in western Alaska. While pickled salmon is occasionally prepared at other points along the coast, no commercial packs have been made elsewhere for a number of years. A production of 829,000 pounds of hard-salt salmon is reported for 1940. This is less than half the average pack of the 1920's. Under market conditions of the past few years, the price of raw fish, salt, barrels, and labor has increased to such an extent that there is no margin of profit except to a few packers with an assures market and with surplus catches which it is not possible to can. It is expected that there will always be a market, though limited, for hard-salt salmon. All five species of salmon are used to some extent in the preparation of hard-salt salmon, but red salmon (*Onchorhynchus nerka*) is the principal species used and yields the highest grade product. A considerable amount of pink salmon (*O. gorbuscha*) and silver or coho salmon (*O. kisutch*) is also salted. While king (*O. tschawytscha*) and chum salmon (*O. keta*) are used rather extensively for noncommercial packs, very little less of these two species is hard-salted commercially. Some pickled salmon bellies are still put up, but the pack is very small. Salted bellies are a choice product but under present law, bellies may only be packed when the rest of the fish is used for food in another way. The major portion of the belly pack is made from pink salmon. Most of the salmon used for brine salting is taken by drift gill nets. In fact, 90 percent of the commercial catch in Western Alaska is made with this type of gear. Some are also taken by staked gill nets and beach or haul seines.

Raw Material Requirements

Quality is an essential consideration in packing hard-salted or pickled salmon. It is important to use only fresh fish, for soft bellies are very evident after salting. If the fish are not in the best condition the bones will come loose from the flesh and stand out, giving the fish a ragged appearance. Flavor, texture, and color are also inferior in the finished product if the salmon are not quite fresh when salted. Salmon approaching the spawning stage cannot be used because of discoloration of the skin and inferior flavor. The use of pews (single-tined fish forks) in handling the raw material is also detrimental. The holes made by the prong cause more rapid spoilage and leave dark streaks in the flesh, detracting from its appearance.

Butchering

When the salmon are landed at the saltery they are washed, slimed, and beheaded. In heading, care is taken not to remove the napes or collar bone. Cutting follows the curve of the gill-flap opening. After heading, the fish pass to the splitters. There are two methods of splitting. In the first, the fish is split along the back, ending with a curving cut near the tail. The abdominal side is left as a solid section. Some two-thirds of the backbone is then taken out, and all viscera, blood, and membranes are scraped away. In the second method the fish is first split along the ventral side, eviscerated, and all membranes are scraped from the belly cavity. In splitting, the neck or head end of the salmon is toward the splitter, who makes a light incision at the neck end, just above the backbone. The whole side is then removed with one sweep of the splitting knife, leaving as little flash as possible along the backbone. The knife is usually held so that the edge of the blade is at a slight downward angle. In removing the second side, two short cuts are first made under the backbone, one on each side, just above the region of the anal opening. Another slight cut is made under the neck end of the backbone, and one sweep of the knife removes the entire backbone and tail. The two cuts are made under the backbone to direct the course of the knife, preventing it from slanting too much in splitting. Slanting causes considerable flesh to be left on the backbone, which is, of course, wasted. Some curers make one or more longitudinal slashes in the flesh so that the salt will "strike" or penetrate more rapidly. The loss of weight in cleaning and splitting averages 25 percent.

Washing

After splitting, the salmon pass to the cleaners. These men scrape out blood clots and kidneys, and remove membranes, loose bones, or other offal. After cleaning, the fish are scrubbed thoroughly, inside and out, care being taken not to injure the flesh. A final cleaning is given in the washing tank and the salmon are then drained thoroughly, preparatory to salting.

Salting

Both round and square salting tanks are used. The tanks have no standard dimensions but the capacity should not be more than 100 barrels of fish. It is preferable to use several smaller tanks rather than one large one. If the tank is too large, pressure on the lower layers of fish is excessive, and the sides of fish in these layers are distorted or otherwise injured. Then, too, a large tank may be more than sufficient for a day's pack.

A thin layer of salt is scattered over the bottom. A layer of fish is then packed in, with flesh side up. No special system is followed in packing, the only rule being that the work must be done neatly, with the layers as level as possible. Each layer of salmon is covered with a thin layer of salt. Care must be taken that each fish is completely covered with salt. The tanks are filled with several layers above the top, to allow for shrinkage, and the top layer is packed with the skin side up. A thicker layer of salt is used to cover the top layer. From 25 to 30 pounds of salt are required to cure 100 pounds of salmon. Several boards are laid across the fish as soon as the tank is filled and these are weighted down with large stones. This keeps the salmon submerged beneath the brine as it forms. The fish are allowed to make their own brine, which is formed as the salt extracts moisture from the flesh. The fish should be covered with brine at all times to prevent "rusting," that is, discoloration caused by oxidation. From 10 to 14 days will be required for this curing process, though the salmon may be left in the tank for a longer period of time. Curers do not agree on the loss of weight in salting. The best estimate at present is that about 15 percent of the moisture content is removed.

Grading and Repacking

The final step is repacking into barrels containing 200 pounds net weight, exclusive of brine. In repacking, the fish are washed in brine and scrubbed well, usually with a stiff brush, though pieces of burlap may be used. All slime, blood clots, excess salt, or other waste material should be removed. The salmon are then sorted into grades: (1) as to species (if more than one species is being cured); (2) color of flesh and skin according to whether the flesh is of good color and skin bright or flesh pale in color, with skin murky or discolored; and (3) as to quality, good or poor; that is, fish which were not strictly fresh when packed and have a characteristic odor and flavor, must be separated from the remainder of the pack.

After sorting, 200 pounds net weight of fish are weight out for each barrel to be packed. A barrel of pickled or hard salmon requires from 40 to 52 red salmon, 25 to 35 coho (medium red or silver salmon), 70 to 80 pink salmon, 10 to 14 king salmon, and 25 to 30 chum salmon. The sides are packed flesh side up, except for the top layer. A liberal sprinkling of salt is scattered at each end, but only a little between the layers. If the fish have been properly cured, from 8 to 10 pounds of salt should be a sufficient amount for repacking a barrel of salmon. After the barrels have been headed they are filled with 100° salinometer brine through the bunghole. One end of the barrel is stencilled with the packers name or brand, the species of salmon, and grade. The barrels should be inspected every few weeks in storage to determine whether they are full of brine and are not damaged or leaking.

Salmon Bellies

A few salteries pack bellies, which are merely the ventral sections, the fattest and choicest portions of the fish. Because of the waste by this method, the preparation of this article was forbidden under the Alaska fishery regulations, unless some economic use was made of the remaining portion of the fish. (Section 8, Act of June 26, 1906, 34 Stat.480; 48 U.S.C. 236).

In preparing salmon bellies, the curer first cuts off the two pectoral fins and then removes the head. Care should be taken to follow the curve of the gill flaps until the backbone is reached, when a straight cut may be made. With smaller salmon, the fish is then turned on its back, a knife inserted vertically in the body, the knife coming out just in front of the vent or anal opening. If made properly, the cut will come close to the upper wall of the belly cavity. With large king (also known as spring or chinook) salmon, it is sometimes necessary to make a cut on one side, then turn the fish over and cut through on the other side. The belly is then laid flat on the cutting table and the membrane at one end cut so that the belly will lie flat.

The bellies are washed thoroughly in clear, cold water, or in iced brine. The salting and curing process is identical with that for hard-salted salmon. Bellies are sent to Seattle or San Francisco markets in barrels holding 200 pounds net, but are usually repacked for distribution to the retailer in kits or tubs of various sizes.

Chapter Fifteen

SPICED AND PICKLED FISH

Spicing and pickling is one of the minor methods of preservation used in North America, but it is very important in Europe and Asia, where a great variety of fishery products prepared with vinegar and spices have wide popularity.

The greatest development in pickled fish preservation in the western world was in Germany. Before World War II that country prepared the best spiced fishery products and also consumed the most products. Many of the published formulae are derived from German sources such as Hoffman (no date), Jacobsen (1926), and Stahmer (1925). There are said to be 200 different recipes for preparing spiced herring, and there are many other species of fish preserved with vinegar, salt, and spices. Other north European countries including Norway, Sweden, Denmark, the U.S.S.R., and the Netherlands produce a wide variety of pickled herring products.

In Japan, the Philippine Islands, and on the coast of China, the pickling of fish is even more important than in Europe. In fact, pickled fish is the most important single source of animal protein in Japan. Here fish are pickled in hot vinegar, either plain or spiced, without salting. Other methods include the pickling of fish in soy sauce, sake, and in a fermenting rice paste.

Spiced and pickled fish have been prepared on a small scale in the United States for many years, but prior to the war, much of the supply was imported. Stevenson (1899) describes the preparation of a number of spiced and pickled products. Until rather recently only a few articles were in demand, on a small scale, and almost entirely in cities with a large foreign population. Now there is a growing demand for pickled fish specialty products in the United States. The Swedish "smorgasbord" restaurants which have spread into a number of American cities have made some of these pickled fish items popular. The cocktail party vogue has also crated some demand for pickled fish. Our tastes, too, are becoming more cosmopolitan.

Pickling with vinegar and spices is a very ancient form of food preservation, going back to prehistoric times. Stevenson (1899) believes that it probably antedates even pickling with salt. It is mentioned frequently in the writings of the Greeks and Romans, as witnessed by the citations of Smidth (1873) and Radcliffe (1921) in their excellent accounts of the fisheries of the ancients. Certain of these fishery products prepared with vinegar and spices were considered great delicacies, selling at such high prices that they were reserved for the banquet tables of the rich. One dish popular in Spain and Latin American republics of Central and South America today is "escabeche." It is prepared by frying fish in oil with bay leaves and spices, then marinating in vinegar and oil. This dish can be traced directly to the Romans, who in turn had it from the Greeks.

Pickling with vinegar was used extensively down through the Middle Ages, especially for fish that were fat and did not cure well by the very crude salting methods of the times. While the product did not keep so long, it was more appetizing than the dried and salted products of the period. Vinegar-pickled fish played a very important part in the food economy of the north European people down through the seventeenth century.

Brine-salted fish is often called "pickled", but this is a misnomer, if the name as applied to other food products is considered. Pickled foods are fermented in the process of manufacture with the formation of organic acids. If the amount of organic acids formed is found not sufficient, more acid may be added in the form of vinegar; or vinegar may be used in the original cure instead of depending on the natural formation of acid. Therefore, only fish preserved with vinegar or vinegar and spices should be considered pickled.

PRESERVATIVE ACTION OF INGREDIENTS

Vinegar differs from salt as a preservative agent in that it does not preserve by osmosis, extracting water from the food, but enters into chemical combination with the product, reducing or inhibiting bacterial activity. The spices used may also have some slight value as a preservative; in fact, through the Middle Ages pepper and other spices were valued as preservatives rather than condiments. Foods of this period were often very highly spiced, and it is probable that spoilage was masked rather than inhibited. Anderson (1925) believes that the preservative action of spices in general insignificant and cites an experiment where an infusion of spice in water developed enough gas in two weeks to blow the cork out of the bottle. The preservative action of vinegar is probably due to the fact that spoilage molds are inhibited and bacteria greatly retarded; if the product has a low pH, that is, if the medium has an acid reaction. The active principle is the acetic acid in vinegar. According to Tressler (1923) an acid content of 15 percent is required to entirely prohibit bacterial action. Since commercial vinegars average 6 percent acetic acid content, and stronger vinegars reduced to this level are as low as 3 percent when used in food preservation, fish products put up in vinegar are only temporarily preserved. However, vinegars containing 3 percent acetic acid will preserve fish for months if the product is held in chill storage.

Tanner (1944) believes that the preservative action of vinegar is probably of little practical importance. He reviews recent work in this field and cites evidence that factors other than pH may be involved. It is suggested that the acetate radical may be more important than pH. The preservative effect of 0.75 percent of acetic acid is stated to be equal to 0.1 percent of sodium benzoate. He also states that in addition to the factors mentioned, others which are present in vinegar may exert in inhibitive action on microorganisms. Vinegar is known to contain propionic acid which has been shown to inhibit the development of molds

REQUIREMENTS FOR INGREDIENTS

Fish used in the vinegar-spice cure must, of course, be of the best quality. Equally important, however, in obtaining a finished product of the highest grade, is the quality of the various accessory ingredients. The flavor, color, texture, and to some extent, the keeping quality are affected by the water, salt, sugar, vinegar, spices, herbs, and other miscellaneous ingredients.

The water must be potable and approved under all sanitary and health codes. Pond or shallow-well water should not be used. "Hard" waters are unsuitable, especially those with a high iron, calcium, or magnesium content. If the water supply is hard, the water must be filtered or a softener used. Brines should be boiled and filtered before use.

The vinegar should be clear, without foreign odors or flavors, and should have a guaranteed acetic acid content. Cider or other fruit vinegars should not be used because their acid content is extremely variable and the fruit esters in the vinegar might give the pickled fish an "off flavor." Distilled vinegar is recommended. Acetic acid diluted to the desired strength may be used. A 120 grain (2 1/2 percent) malt vinegar is favored by some commercial packers. Vinegar should be titrated regularly to determine whether it has the required acid content. Wine may be used as a flavoring agent, as in a spiced wine sauce added to the pickled fish, but wine or wine vinegar should not be used in the original pickle cure.

The quality of the salt is also important. A high grade of purity is required. European curers prefer a Liverpool fishery salt. The salt must be as free as possible from calcium and magnesium compounds, as these impurities give a bitter flavor to the cured product. A coarse, half-ground salt will pit the surface of the pickled fish. A finely ground cooking salt, guaranteed to contain less than one percent impurities and preferably mined, is recommended.

Experienced packers of spiced fish prefer a good grade of cane sugar. Some use is made of corn sugar but the quantity must be adjusted to obtain the same results with the standard formula. It is not a completely satisfactory substitute to date. This also applied to beet sugar. Saccharine has been used but is not recommended. It may be used in some countries but not in the United States where its use is prohibited by the Food and Drug Administration.

Spices should be bought whole and on the basis of chemical and microscopic examination. Buying specifications should require that the spices be fresh and of high grade of purity. Ground spices should not be purchased as they may be readily adulterated with other plant material. Mixtures of whole spices may be composed of old or inferior spices. As a rule the curer should buy fresh whole spices, singly, and make up his own mixture. Spices should not be held much more than a year.

Essential spice oils should meet the requirements of the United States Pharmacopoeia. Both spices and oils should be bought only from well-known, reputable firms specializing in these products. "Bargain" spices are usually ruinous to quality products.

HERRING

Herring is the most important pickled fish product. Some of the more common pickled-herring products are appetitsild, Bismarck herring, cut spiced herring, gabbelbissen, gaffelbiter, Kaiser Friedrich herring, potted herring, rollmops, Russian sardines, gewurz herring, delicatessild, sur-sild, and smorgaas-sild. The names indicate that most of the spiced herring products prepared in the United States are of foreign origin. The composition of the sauce is the principal point of difference between many of these products. Kaiser Friedrich herring, for instance, are Bismarck herring in mustard sauce, while rollmops are Bismarck-herring fillets rolled around a piece of dill pickle. Bismarck herring are boned, with sides still joined, and packed in vinegar and spices.

Spiced herring packed in the United States is prepared entirely from fish cured during the fishing season, and made into spiced herring products as the market requires. Herring may be specially cured for this purpose, although Scotch-cured or Labrador salt herring may be substituted. Herring not specially cured for spicing is reported to have a shorter period of preservation, to be darker in color, lacking in flavor and tougher and more fibrous in texture. One of the most common methods for home preparation of salt herring is to pickle it with vinegar and spices. At one time salt herring were used extensively for commercial vinegar-pickled herring but such raw material is used today only when the stock of specially cured herring is exhausted, or in places where they cannot be obtained.

Most of the specially-cured herring are prepared from alewives or river herring (*Pomolobus pseudoharengus*) in the Chesapeake Bay area. Some are made from Atlantic herring (*Clupea harengus*) at Eastport, Maine, and a few neighboring areas. The Pacific herring (*C. pallassii*) is found from Puget Sound to western Alaska. A variety of curing methods are used, of which the following is typical: The herring are cut or dressed as described in the salting of alewives. The cut herring are cleaned thoroughly, with special attention to removal of the kidney, which is the dark streak along the backbone. The fish are rinsed in fresh water and placed in a curing tank where they are covered with a brine testing 80° to 90° salinometer that contains 120-grain distilled vinegar amounting to an acidity of about 2 1/2 percent. The fish are left in this brine until the salt has struck through, or completely penetrated the flesh. The fish, however, must be removed before the skin starts to wrinkle or lose color. The length of cure depends on the judgment of the curer and varies with the temperature conditions, freshness and size of the fish. The average length of cure is reported to be 5 days. Various sources of information give curing times varying from 3 to 7 days.

When the herring are sufficiently cured they are placed in barrels. These are often second-hand, previously used for soda fountain syrup. As a rule, no attempt is made to pack in regular layers. The herring are simply shoveled in until the barrel will hold no more. The barrels are headed, filled with a salt-vinegar brine testing 70° salinometer, and shipped to marketing centers such as Chicago or New York for final manufacture. There the herring are repacked in kegs which are then filled with a solution of distilled vinegar diluted with water to a 3 percent acidity, and containing sufficient salt to test 35° salinometer. Before the fish are repacked they must be cut into fillets or the backbone may be removed with the fish left otherwise whole. The kegs are then put into cold storage at 34° F. to be held until required.

The final process of manufacture is begun by soaking the herring in a tank of cold water from 8 to 10 hours. They are then removed, drained and placed in a solution of vinegar, salt, and water for 72 hours. The solution is made in the proportion of 1 gallon of 6 percent white distilled vinegar to 1 gallon of water and one pound of salt. The fish should be well covered with the solution. They are then made into cut spiced herring, rollmops, or Bismarck herring.

Cut Spiced Herring I

The vinegar-salt cured herring are cut across the body in pieces 1 to 2 inches long. The sliced herring are then packed in 8-ounce tumblers or in 16- or 32-ounce glass jars, with whole mixed spices. The amount of spices added to each container is approximately one teaspoon to an 8-ounce tumbler, 2 teaspoons to a 16-ounce jar and 1 tablespoon to a 32-ounce jar. A slice or two of onion, a slice of lemon, or a strip of canned pimiento, and a bay leaf or two may be placed around the sides for ornament, depending on the preference of the individual packer. Each container is then filled with vinegar diluted to 2 1/2 percent acidity, containing 1/2 pound of sugar,, 1/4 pound of salt, 10 drops oil of cloves, 10 drops oil of allspice, and 10 drops oil of cardamom per gallon of solution. The spice oils are usually added to the sugar before dissolving it in solution to distribute the spice flavor more evenly.

The amount and variety of spice flavors may be altered to suit the taste of the packer or his market. The formula given is typical but is not claimed to be standard. The jars are vacuum sealed, wiped clean, labeled, and packed one or two dozen jars to the fiberboard carton. The length of preservation depends on the care in manufacture and the temperature of the storage. If held under refrigeration at 40° F. this product should remain in good condition at least 6 months. Exposure to light causes the herring to deteriorate more rapidly, even if held under refrigeration, as in a refrigerated showcase.

Cut Spiced Herring II

The cut pieces of herring are packed in wooden tubs holding 10 or 20 pounds, or if they are to be repacked in individual glass containers, in kegs holding 100 pounds. A few spices, a bay leaf or two, and several slices of onion are placed in the bottom of the container, then a layer of cut herring, on which are laid spices and onion. This is repeated until the tub or keg is filled. The fish are covered with vinegar diluted with water in which the sugar is dissolved. The containers are then stored at 40° F. to cure for 10 days. At the end of this time if the fish are to be repacked, they are removed from storage and filled into 8-, 16-, and 32-ounce glass jars. The vinegar used in curing may be strained and re-used but some prefer to use fresh vinegar diluted to three percent acidity. A few of the spices, bay leaves, and a little chopped onion are placed in each jar.

The quantities given in the formula below are sufficient for 10 pounds of cleaned herring. Whole spices are used in all recipes, unless otherwise specified.

There are other recipes for the preparation of cut spiced herring. Cut spiced herring in tubs usually go to delicatessen shops or other wholesale outlets.

10 lb. salt herring
2 oz. mustard seed
1 oz. bay leaves
1 oz. allspice
1 oz. black peppers
1 oz. white peppers
1/2 oz. cloves
2 1/2 oz. sugar
4 oz. onions, sliced
2 qt. water
2 qt. vinegar (6% acidity)
1 oz. red (chili) peppers

Rollmops

The vinegar-cured fillets are wrapped around a piece of dill pickle or a pickled onion. The rolls are fastened with wooden toothpicks, cured several days in a spiced-vinegar sauce, then packed in glass containers, generally of the same sizes used for cut spiced herring. Anderson (1925) made a comprehensive study of packing rollmops. The formula given here is the one recommended by him:

10 lb. salt herring	1 /16 oz. powdered nutmeg
4 oz. chopped onions	1/16 oz. cracked cinnamon
2 1/2 oz. sugar	1/4 oz. mustard seed
1/4 oz. whole cloves	1/16 oz. cracked ginger
1/4 oz. chili peppers	1/8 oz. bay leaves
1/16 oz bay leaves	1/16 oz. crushed cardamom
1/8 oz. whole black peppers	2 qt. distilled vinegar (5 %)
1/8 oz. whole white peppers	1/4 oz. whole allspice
8 qt. dill pickles	

"Put the bay leaves and chili peppers in a small cloth bag so they can be easily separated for later use. Place this bag together with the balance of the spices and three-fourths of a quart of vinegar into a covered receptacle. Bring to boil and allow to simmer for one and one half to two hours. Violent boiling causes loss of the volatile acetic acid. A very simple way is to put the spices in a common fruit jar and place in boiling water for two hours. Allow to stand 1 to 2 weeks after boiling to insure still greater extraction of the spicing materials. Remove the chili peppers and bay leaves which are to be used for decorative purposes. Strain the pickle through a cloth bag to remove the spices. These should be well mixed, ready for adding to the jars before packing. Slightly less than three-fourths of a quart of pickle will be obtained.

Preparation of the fish: Remove heads, scale and wash. Split into two fillets and trim. Freshen two and one-half hours in running water, then drain. Ten pounds of medium size herring should give about 6 pounds drained weight.

Preliminary vinegar-cure: Pack the fillets skin down in a stone crock. Cover with one and one-fourth quarts of vinegar. If necessary, put a light weight on top to keep the fillets well covered. Allow to cure in a cool place for 40-48 hours. Remove and drain. The vinegar should now test about 2 percent acid and show a salinometer reading of about 30°. The fish has absorbed much of the acid and has lost some salt.

Packing: Cut each dill pickle lengthwise into four parts, then each of these across the center, making eight pieces in all. Roll the fillets around a piece of pickle and fasten with a fresh clove. A clove serves the purpose just as well as a toothpick and adds to the attractiveness of the pack.

Place one teaspoonful of mixed used spices on the bottom of the jar, then pack the fish. With a medium sized herring, three rolls will pack nicely into a No. 306 jar (six fluid ounces capacity) if placed on end. Decorate around the sides with a couple of chili peppers and a bay leaf. Add sufficient pickle to fill from 25 to 35 cc. (this is about equivalent to two level tablespoonfuls). The net weight should be five and one-half ounces or over. Seal the jars immediately after packing. Vacuum sealing is preferable.

Store in a cool place. Cold storage at about 35° F. is advisable to insure longest preservation."

If vinegar-salt-cured herring are used, the preparation and preliminary vinegar-cure steps will be unnecessary. It is believed that a better product will be obtained if the vinegar-salt-cure herring are used. Test packs, by the writer, of Anderson's formula indicate that the spice-vinegar sauce should be diluted to 3 percent acidity, that the rollmops should be cured in the spice sauce for 10 days, then repacked in jars with a few spices, and the jars should be filled with fresh 3-percent vinegar with 2 tablespoons sugar and 1 of salt to the quart. Store at 34° to 40° F.

Bismarck Herring

The vinegar-salt-cured fish are treated as in the preparation of cut spiced herring up to the point of slicing. Instead of slicing, the backbone is removed, but the two sides are left joined together along the back and trimmed of thin ragged edges of belly flesh. The boned and trimmed fish are packed with vinegar and spice oils or spices as described for the preparation of cut spiced herring. They are left to cure at 40° F. for 10 days. At the end of this time, the fish are packed vertically in glass containers of sizes such as 8-ounce tumblers, and 16- or 32-ounce glass jars. A few fresh spices and a bay leaf with a slice of lemon or other similar decorative agents to suit the preference of the packer are placed around the sides of the jars. The jars are then filled with distilled vinegar, diluted to 3-percent acidity. The containers are sealed, cleaned, and packed into cartons.

Gaffelbiter

Take fat mild-cure salt herring, cut into fillets, and skin. The fillets are then freshened in running water 2 to 3 hours, depending on the size of the herring, whether mild or heavy cure, and local market preferences. The fillets are then cut into sections about one inch in length and packed into tubs or small crocks, with chopped onion, bay leaves, whole black peppers, whole cloves, and mustard seed, mixed with the fish. The herring are then covered with distilled vinegar, from 3 to 5 percent acidity, and left to cure in cool storage (about 40° F.) for 48 hours. They are then packed in glass containers, filled with fresh vinegar or the vinegar used in curing, but strained before reuse. The jars are sealed, wiped, and packed into cartons.

This product differs from Gaffelbiter as prepared in the Scandinavian countries. It is descriptive of the method as developed in the United States for the Scandinavian-American market. The formula was obtained from a Scandinavian-American fish curer. The ingredients are given below:

16 lb. mild-cure herring	1/4 oz. whole black peppers
1 qt. 6% distilled vinegar	1/4 oz. whole white peppers
1 qt. water	1/4 oz. whole cloves
1/4 oz. mustard seed	8 oz. chopped onions
1/8 oz. bay leaves	

Gabelbissen

Fresh fat herring is used for gabelbissen. The round herring are first cured for 30 to 40 hours in a brine testing 90° salinometer. In some instances, however, this preliminary brine cure is omitted and the fresh fish are packed directly into barrels. A special curing mixture is scattered on the bottom of the barrel between the fish and over each layer. The herring are packed much as described for the packing of whole salt herring, with bellies straight up and tails overlapping. The layers of fish are packed in rather loosely. The barrels are headed up and put in cold storage at about 40° F., where they are held for several months to cure and ripen.

At the end of about 3 months the herring are removed from storage, drained well, headed, boned, and skinned. The fillets may be cut into sections and packed in glass containers, or whole fillets may be packed in oval or oblong flat cans. A bay leaf and a thin slice of lemon may be laid in each can. The containers may be filled up with the original curing brine diluted one-half with distilled vinegar or they may be packed in wine sauce, dill sauce, or curry sauce. The containers are sealed, cleaned, packed in cartons and held under refrigeration until sold to the retail consumer. The curing mixture is as follows:

220 lb. herring	4 oz. cardamom
22 to 35 lb. salt	2 oz. ginger
4 lb. sugar	6 oz hops
2.2 lb. black peppers	2 oz. cloves
2.2 lb. white peppers	2 oz. cinnamon
18 oz. allspice	3 oz. saltpeter
11 oz. coriander	

The spices are ground coarsely, at the time the mixture is made up, and mixed thoroughly with the salt, sugar, and saltpeter.

Matjeshering

This is another spiced-herring product considered very choice in northern Europe. It is prepared from fresh, full herring, that is, herring with milt or roe. The fish are first washed and scaled. Then the gills are removed and the intestines are pulled out through the gill opening so that the throat or belly wall are not cut open. The fish are soaked from 12 to 18 hours in a 7 percent solution of white wine vinegar. They must be taken from this solution before the skin becomes soft and flabby.

They are wiped dry and rolled in a curing mixture consisting of 2.2 pounds salt, 1.1 pounds brown sugar, and 4 ounces saltpeter. This amount is supposed to be sufficient for the unit quantity, a small keg of 75 herring. The next step is to pack the herring in small kegs, with the backs up and the layers straight. Some of the curing mixture is scattered between the fish as they are packed and some is sprinkled over each layer.

After 24 to 48 hours the fish are repacked, using the original brine. If enough brine has not formed, sufficient additional brine to cover the fish is made from one part of the salt-mixture to four parts water. This is boiled, cooled, and filtered before using. The kegs are stored at 40° F. for at least one month or until the product is considered ready for use.

Gewurzhering

This product is prepared in the same way as matjeshering, but spices are added to the curing mixture.

75 herring	1 oz. ginger
4 oz. black peppers	1 oz. cloves
2 oz. chili peppers	1 oz. cinnamon
2 oz. allspice	1 oz. nutmeg
1 oz. coriander	

The spices are ground and added to the salt-sugar mixture when it is made up.

Russian Sardines I

This is a favorite product in northern Europe, where large quantities are prepared. It has been packed in the United States for about 80 years. Production has been small of recent years since this product requires small herring of the size most desirable for Maine sardine canning. There are several different methods of cure, or which two are given here.

Wash and scale 10 pounds of small, fresh round herring (from 5 to 7 inches). Remove the gills and as much of the intestines as possible by pulling them out through the gill flap without tearing the throat or belly. Rinse again, drain, and pack in a large crock or tub. Cover with 3 parts of distilled vinegar and 1 part water. For this amount of fish, 3 pints of 6 percent distilled vinegar and 1 pint of water will be required. Allowthe herring to stand for about 12 hours.

10 lb. fresh herring	1/2 oz. bay leaves
2 lb. fine salt	1/2 oz. cloves
1 lb. powdered sugar	1/2 oz. ginger
1 oz. allspice	1/2 oz. hops
1 oz. black peppers	1/2 oz. nutmeg
1/2 oz. saltpeter	

The spices should be finely ground just before use and blended with the other ingredients.

After the fish have drained, they are dredged in the spice mixture and packed in crocks or tubs, bellies up. A small additional amount of the mixture may be scattered between each layer. The layers should be packed at right angles to each other with the top layer packed backs up. The balance of the spice-curing mixture is scattered over the top layer and the fish are weighted down so that they will be entirely covered when the brine forms. Some packers also scatter chopped onions, ground or sliced horseradish, and capers between each layer. The amounts of these ingredients required for 10 pounds of small herring are: 1/2 pound onions, 1/4 pound horseradish, and a small bottle of capers (about 2 1/2 ounces). The fish are held in cool storage for at least two weeks before repacking in small kegs holding about 7 pounds, or in glass jars. The brine formed in curing is filtered and poured into the containers before closing.

Russian Sardines II

Fresh small herring (5 to 7 in.) are packed in 100° salinometer brine as soon as possible after catching. They are left in the brine about 10 days until they are thoroughly salt-cured or struck through, the exact time depending on the size of the fish and the weather. The salted fish are then headed, pulling out the viscera with the same stroke of the knife without tearing the belly open. They are then washed in clean water and placed on wire trays for draining. After the herring have drained for several hours they are taken to packing tables where they are sorted according to size. Each size is packed separately in small kegs holding about 7 pounds each. A scattering of spices and flavoring ingredients is laid in the bottom of the keg. A layer of herring is placed with backs up. The layer is then pressed down slightly. A thin layer of spices and a little vinegar is then added. The process is repeated until the keg is filled. As much vinegar as the keg will hold is then poured in and the container is headed up. The fish require some time to season before they are put on the market. In summer they are ready for sale in 4 or 5 days but in winter 3 or 4 weeks may be required. If the product is held in refrigeration at about 40° F. it will remain in good condition for a year. Stevenson (1899) gives the following ingredients as sufficient to cure 120 pounds of herring:

120 lb. fresh herring	1/2 lb. ginger
2 gal. vinegar	4 lb. sliced onion
1.5 lb. allspice	2 lb. horseradish
1 lb. bay leaves	1/2 lb. chili peppers
1/2 lb. cloves	1/2 lb. coriander seed
	2 1/2 oz. capers

Other spices may be used to suit the preference of the market and packer.

Potted Herring

This is a pickled herring dish greatly appreciated in the British Isles. Fresh herring are always used. The herring are first scaled, headed, split down the belly, and washed thoroughly. After draining, the inside of each fish is rubbed with a mixture of black pepper and fine salt. The herring are then laid in layers in a baking dish with a few whole cloves and bay leaves scattered over the layer of fish. When the dish is filled the fish are half-covered with vinegar and the dish is baked in a moderate oven (about 350° F.). This product will keep about 2 weeks at ordinary temperatures (Jarvis 1943a). Additional preparation in a somewhat different form is described in another section of this report. Small mackerel are also cured by this method.

Scandinavian Anchovies I

The true anchovy is not used in the preparation of this product. The brisling or sprat (*Clupea sprattus*) is used. In contrast to the Spanish or Portugese anchovies (p. 157, ibid), Norwegian or Swedish anchovies are flavored with spices and the curing is not primarily a fermenting process. Various spice mixtures are used and the methods differ in detail but the following is typical: From 25 to 30 pounds of brislings are cured for 12 hours in a brine made of 4 1/2 pounds of Liverpool salt and 7 quarts of water. At the end of the salting period the fish are laid on a wire screen to drain. The spice-salt mixture is then made up, with all spices well pulverized and the ingredients thoroughly blended. The sprats are stirred in well in half of the spice mixture and packed in a large container where they are left to cure for 14 days. They are then repacked in individual containers in layers, bellies up. Some of the remaining spice mixture is scattered between each layer, with pieces of chopped bay and cherry leaves. On the bottom and top of each container two whole bay leaves are laid. The brine formed in the original spice cure is filtered and used to fill the smaller containers after packing. During the first few days after the containers are closed they must be rolled about and inverted at least every other day (Hoffman; undated).

The ingredients for the spice mixture and the amounts for the quantity of fish given here, are as follows:

2 1/4 lb. Lunenberg salt	3 oz. sugar
1/2 oz. cloves	3 oz. black peppers
1/2 oz. nutmeg	3 oz. allspice
1/2 oz. cayenne	

Tin containers are preferred to wooden kegs, which are often leaky and the airtight seal of tin containers permits a longer period of preservation.

Scandinavian Anchovies II

Fresh brisling are placed in a strong salt brine from 12 to 24 hours. They are drained on a screen, and are packed in layers in small kegs after being rolled in a spice-curing mixture. Some of this mixture is scattered between the layers of fish. At the top, bottom, and in the middle of the keg, several bay leaves are laid. The kegs are packed tightly and are rolled about or inverted for 14 days. The anchovies may be repacked in tins in 14 days in summer or after 4 to 8 weeks in winter.

40 lb. brisling	7 oz. sugar
2 1/4 lb. Luneberg salt	1 1/8 oz. cloves
1 1/8 oz. nutmeg	7 oz. black pepper
1 1/8 oz. Spanish hops	7 oz. allspice

The spices are ground and mixed with the salt.

Scandinavian Anchovies III

A third formula, from German sources, is as follows: The brisling are brine salted in Norway in barrels holding about 100 kilograms (220 pounds). When the brisling are to be manufactured into anchovies, the barrels are taken out of cold storage and the fish washed thoroughly in a light brine testing 40° salinometer. After draining, the fish are packed loosely in new barrels with some of the spice mixture scattered between each layer of fish. The brine used for washing and the original brine are filtered and poured into the barrels after filling. The barrels are placed in cool storage for several months for the fish to ripen or acquire an aromatic flavor. The barrels should be rolled about daily, or at the least, at intervals of two or three days. When the fish have completely absorbed the spice flavor they are repacked in small individual containers, small kegs holding about 7 pounds, cans holding from 2 to 5 pounds, and glass jars. The brine used in curing is filtered and filled into the containers when they have been packed with fish. In summer 0.5 percent benzoic acid may be added to the brine.

The spice mixture, in quantity for one barrel of original weight, is as follows:

2 lb. black peppers	6 oz. Spanish hops
1 lb. allspice	2 oz. mace
1 1/2 lb. sugar (best raw)	2 oz. cloves
2 oz. cinnamon	1 lb. saltpeter
2 oz. ginger	1 lb. bay leaves

Appetitsild

The largest size Norwegian style anchovies cured by the method just described are selected and after heading are cut into fillets, removing the backbone. The fillets are packed into oblong flat cans, flesh side up, with a bay leaf and a thin slice of lemon in each can. Filtered curing brine is poured into each can, which is sealed immediately. The cans are not processed.

Small fat herring may also be used in the preparation of anchovies or appetitsild. According to Jacobsen (1926), brisling, when cured as anchovies, should have a salt content of 10 percent and a moisture content of 40 percent. Fat herring when cured as anchovies should have a salt content of 14 percent and a moisture content of 45 or 50 percent.

The blend of spices in anchovy curing is considered very important. The flavor must be a blend, with no single spice predominating. The best storage temperature is considered to be about 40° F.

Herring in Sour Cream Sauce

Mild-cure salt (Holland style) herring are filleted, saving the milts. The fillets are soaked in cold water for two hours. The milts are rubbed through a fine sieve. The fillets are drained of surplus moisture. The vinegar, wine, and spices are boiled together for a few minutes, then cooled, and after removing the spices, stirred into the sour cream, sweet cream, and milts. The fillets are packed in a container with sliced onions, then covered with the sauce. The whole is allowed to marinate in a cool place one week. The fish are then packed in glass jars, which are then filled with the sauce.

1 gal. keg Holland style herring	1/2 pt. distilled vinegar
1/2 oz. mixed spices	1 pt. white wine, dry
1 pt. sour cream	2 cups onions, sliced thin
	1 pt. sweet cream

Herring in Wine Sauce

Wine sauce formulas in general are like the standard spice sauce formulas for herring, with the exception that the amount of vinegar is reduced to one-half or three-fourths and wine used instead. A dry white, or burgandy-type red wine must be used. Sweet wines are not suitable. A sample wine sauce is given as follows:

1 qt. white wine	1/8 oz. whole white peppers
1 pt. distilled vinegar	1/4 oz. whole allspice
4 oz. chopped onions	2 1/2 oz. sugar
1/16 oz. ground nutmeg	1/4 oz. whole cloves
1/16 oz. cracked cinnamon	1/4 oz. chili peppers
1/16 oz. cracked ginger	1/8 oz. bay leaves
1/8 oz. whole black peppers	1/16 oz. crushed cardamom

First, the vinegar and wine are poured into a large covered jar, and the onions, sugar, and spices added. The jar is placed in a pan of boiling water for two hours. The jar is allowed to stand overnight. When the sauce is ready to be used it is strained to remove the spices.

If vinegar-salt-cured herring are being used, the herring fillets are cut in pieces of suitable size, rinsed in fresh water, drained, and packed in glass jars with a few fresh spices, such as bay leaves, chili peppers, and a slice of lemon. The jars are filled with wine sauce, then sealed. The amount of sauce above is enough for about 10 pounds of fish.

If ordinary salt fish are used, they are filleted and freshened in water. The drained fillets are packed in a stoneware crock, covered with 1 quart distilled vinegar 3 percent acidity (for 10 lb. herring) and allowed to stand 48 hours. The fillets are then cut in cubes and packed in jars, which are filled with wine sauce.

SALMON

Pickled salmon is packed commercially only in a few localities in northern Europe. It is prepared to some extent in the United States but only for local use. Properly made, it is an appetizing article of food, with a length of preservation about equal to other pickled fish. Commercial formulae are derived from old, home recipes. Representative samples are given here.

Formula No. 1

Ten pounds of salmon are cut into individual serving portions. The pieces are washed well in cold water then drained and dredged in fine salt. The salt is drained off after 30 minutes, and the salmon simmered slowly until done. The warm fish are placed in an earthenware crock and covered with a vinegar spice sauce made as follows:

1 qt. distilled vinegar	1/2 Tbs. bay leaves
1 Tbs. white peppers	1 qt. water
1 Tbs. mustard seed	1/2 cup olive oil
1/2 Tbs. cloves	1 cup sliced onions
1/2 Tbs. black peppers	

Cook the onions in olive oil slowly until they are yellow and soft. Add the rest of the ingredients and simmer the whole gently for 45 minutes. Allow the sauce to cool, then pour it over the fish, making sure that all of the pieces are covered. Allow the fish to stand for 48 hours then repack it in pint jars, with a slice of lemon and onion and one bay leaf around the sides for decoration. Filter the spice sauce, then pour it over the top of the fish until the container is filled, then seal. This product should be held at 40° F. The method may also be used for shad and other large fish.

Formula No. 2

This formula was obtained from a Scandinavian source. The salmon must be absolutely fresh. After it has been well washed, the backbone is removed and the sides trimmed of the very thin belly flesh. It is then cut in very small pieces of about one-fourth pound each. These are simmered in well salted water until they are done but not soft. A sauce is made of the water used in cooking (first filtered), equal parts of white wine and vinegar. The pieces of salmon are packed in wide-mouth glass jars with two or three thin slices of lemon, two bay leaves, four cloves, and four whole black peppers to each jar. The sauce is poured over the fish while still warm, so that all pieces are completely covered.

A three-fourths inch layer if olive oil is poured in the top of each jar. The jars are sealed and then stored in a dry, cool place.

Some mild-cured salmon is cut into two inch cubes, freshened in cold water, and packed in spiced vinegar sauce or wine sauce.

MACKEREL

Fillets

This formula, developed for mackerel, may be used for other fish. It has been obtained from German sources. The fish is cleaned and washed thoroughly, then cut into fillets, removing the backbone. Divide the fillets into 2-inch lengths and dredge in fine salt. Pick up with as much salt as will cling to the flesh and pack in a crock or tub. Let the fish stand for one to two hours, then rinse in fresh water. Cook the vinegar, water and other ingredients slowly and gently for 10 minutes after reaching boiling point. Add the fish and cook slowly for 10 minutes longer, counting from the time at which the solution begins to boil after the fish has been put in. Remove the fish and allow the pieces to drain, then pack them in sterilized jars, adding some chopped onion, a bay leaf, a few spices, and a slice of lemon to each jar. Strain the spice vinegar sauce and bring to a boil. Fill the containers with hot sauce and seal immediately. Store in a cool, dry place.

10 lb. mackerel fillets	2 oz. sugar
2 qt. distilled vinegar	1 Tbs. allspice
3 pt. water	1 Tbs. cloves
2 cups chopped onion	1 Tbs. black peppers
1 clove garlic, chopped	1 Tbs. bay leaves
	1 Tbs. crushed nutmeg

HADDOCK

Fillets

This formula is designed for haddock but may be used for almost any other large fish. The haddock are cleaned carefully, skinned, and cut into fillets, removing the backbone. The fillets are cut in pieces about 2 inches square. These are washed, drained, then put in a tub and covered with a 90° to 100° salinometer brine. They are left in brine from 1 to 6 hours, depending on individual preference. The fish are rinsed in fresh water, then packed in a large crock as follows: A few spices and pieces of sliced onion are scattered on the bottom. Then, pack in a layer of fish, cover with a scattering of spices and sliced onion, and continue until all the fish are packed. The fish are covered with a solution of two parts vinegar and one part water, adding a small piece of alum about the size of a walnut. Boil slowly until the fish may be pierced easily with a fork. After cooling, the product is packed in glass containers, adding a few fresh spices, a bay leaf, and a slice of lemon around the side of the jar for decoration. A few slices of onion may also be packed with the fish. Strain the vinegar sauce, heat it, and pour over the fish until the top is well covered. Seal the containers immediately. For maximum preservation, store this product under refrigeration. Amounts given below should be sufficient for 10 pounds of fish ready for pickling:

10 lb. haddock fillets	1/2 oz. chili (red) poppers
2 qt. distilled vinegar	1/2 oz. allspice
1 qt. water	1/4 oz. cloves
1/2 oz. white peppers	1/4 oz. mustard seed
1/4 oz. bay leaves	1/4 lb. sliced onions

STURGEON

This method is used for sturgeon, pike, pickerel, salmon, herring, trout, and other fish. It is a commercial formula of German origin. Ten pounds of fish are washed well and cut in small individual small serving-size portions, from 2 to 4 ounces each. These pieces are dredged in fine salt and left from 1 to 3 hours. The salt is rinsed off, the pieces are dried, and brushed with good cooking oil. They are laid on a grill and boiled over a hot fire until both sides are a light brown. They should be brushed with cooking oil during the process. The fish is allowed to cool, then packed in glass containers with a slice or two of lemon, bay leaves, onion, and a scattering of rosemary, whole black peppers, and whole cloves between the layers of fish. The jars are filled with a marinade made of white wine, vinegar and water, sealed immediately, and stored in a dry, cool place.

10 lb. sturgeon	1 qt. white wine
1/2 lb. lemon slices	1 qt. distilled vinegar
1/2 lb. sliced onions	1 pt. water
1 oz. whole black peppers	1/4 oz. rosemary
1/2 oz. whole cloves	

Thyme may be substituted for rosemary, and the spice combination may be otherwise altered to suit the individual preference.

Escabeche

This recipe is of Spanish origin. Escabeche was brought to Spain by the Romans, who, in turn, had it from the Greeks. It is probably the most popular Spanish method of preserving fish and is found in all Latin-American countries. Though there are many local variations, all are founded on the basic recipe. Mackerel, kingfish (king mackerel), tuna, and corvina are the fish most used for escabeche. Almost any fish may be used although soft-flesh fish do not make so good a product.

10 lb. fish	1 pt. olive oil
1 qt. distilled vinegar	1 Tbs. red (chili) peppers
2 Tbs. bay leaves	1/2 Tbs. cumin seed
1 Tbs. black peppers, whole	1/2 Tbs. marjoram

Cut the fish into small serving portions. Wash the pieces thoroughly, drain and place in a 90° salinometer brine for 1/2 hour. Then wipe the fish dry. Pour the olive oil into a frying pan, together with a clove of minced garlic, half a dozen bay leaves, and a few red peppers; then heat the pieces until they are light brown in color and lay aside to cool.

Add onions to the oil and cook them until they are yellow. Then add black peppers, cumin seed, marjoram, and vinegar. Cook slowly for 15 to 30 minutes and cool. When the fish are cold, pack into sterilized jars with the rest of the bay leaves and red peppers. Fill the jars with sauce and close immediately. This preparation may be used after 24 hours but it tends to improve with storage. It will keep about 3 weeks in summer; much longer if stored in a cool place.

Pickled Eels

This dish is a favorite in northern Europe, from the British Isles to Sweden. Clean and skin the eels and cut them into pieces about 3/4-inch thick. Wash and drain the pieces, then dredge with fine salt and allow to stand from 30 minutes to 1 hour. Rinse off the salt, wipe the pieces dry, and rub them with a cut clove of garlic. Brush the eel with melted butter and broil until both sides are a light brown. As an alternative, pieces may be sautéed in olive oil or other good salad oil. Place the pieces of cooked eel on absorbent paper. When the pieces are cool, pack them in layers in a crock with a scattering of sliced onion, allspice, bay leaves, mustard seed, whole cloves, peppers, and mace between the layers of fish.

Weight the mixture down to keep it compressed. Cover the fish with a cold vinegar sauce made of vinegar, water, onions, and a few bay leaves cooked for 15 to 20 minutes. After standing for 48 hours in a cool place pack the eels into glass tumblers with a thin slice of lemon, a bay leaf, a slice of onion, and a few fresh whole spices for decoration. Fill the tumblers with sauce used in curing, which has been filtered. Seal the containers immediately. Store in a cool, dry place. This article remains in good condition for a considerable period of time.

OYSTERS

In the eighteenth and nineteenth centuries pickled oysters were prepared commercially over most of the Atlantic coast area. According to Stevenson (1899) pickled oysters were consumed extensively around New York during the Christmas holidays. They are not nearly so popular now and are usually prepared only for special orders. Some pickled oysters are prepared in Virginia for local consumption. Various formulae are available, some from traditional colonial sources, while others may be found in French or other continental cookbooks. Two typical formulae are given here:

1 gal. shucked oysters	1 Tbs. crushed fennel
3 pt. oyster liquor	1 Tbs. crushed allspice
1 pt. distilled vinegar	1 Tbs. crushed black peppers
1 pt. white wine, dry	1 Tbs. crushed cloves
2 Tbs. ground onion	1 Tbs. crushed stick cinnamon
2 cloves garlic, crushed	1/4 Tbs. crushed mace
2 Tbs. bay leaves, crushed	1/4 Tbs. crushed thyme
1 Tbs. chopped parsley stems	

Virginia Pickled Oysters I

Open one gallon of oysters, saving the liquor. Strain the liquor and add sufficient salted water to bring the amount up to 3 pints. Simmer the mixture gently over a low flame. When the liquor is near the boiling point, add the oysters a few at a time, cooking until the "fringe" curls. The oysters are then removed from the liquor and set aside to cool. Make a sauce of vinegar, white wine, bay leaves, onion, garlic, parsley, fennel, thyme, cloves, black pepper, allspice, cinnamon, and mace. Add this sauce to the oyster liquor and simmer 30 to 45 minutes. When it is cool, pack the oysters in glass jars with a bay leaf, slice of lemon, and a few fresh spices in each jar. Strain the liquor and when it is cool pour it into jars, seal immediately, and store in a cool, dark place. The oysters are ready for use in 10 to 14 days.

1 gal. shucked oysters	1/2 oz. allspice, whole
2 qt. oyster liquor	1/2 oz. black peppers, whole
1 qt. vinegar	1 blade mace
1/2 oz. cloves, whole	

Pickled Oysters II

Blanch one gallon of freshly opened oysters until the fringe curls. Remove the oysters and set them out to cool. Bring the liquor used in blanching to the boiling point, then set aside to cool. At the same time cook the vinegar and spices together slowly, after which strain out the spices. Combine the oyster liquor and spiced vinegar. Pack the oysters in glass tumblers with a bay leaf and a thin slice of lemon in each. When the sauce is cool, strain it and pour over oysters until the containers are filled. Seal the tumblers immediately and store in a cool, dark place.

Mussels

This formula may also be used in pickling clams and oysters. Scrub the shells well and steam just enough to open the shells. Save the liquor or nectar. Remove the meats from the shells, cutting out the byssus or beard. Cool meats and nectar separately. When cool, pack the meats in sterilized glass jars with a bay leaf or two, a few whole cloves, and a thin slice of lemon to each jar.

Strain the liquor obtained in steaming the shellfish. To each quart of liquor add one-half pint of distilled vinegar, one-half tablespoon each of allspice, cloves, and red peppers, with one-quarter teaspoon of cracked whole mace. Some recipes for pickled mussels call for white wine or wine vinegar instead of distilled vinegar. The amount of wine or vinegar is a matter of personal taste. Simmer for 45 minutes. When the sauce is cool, pour into the jars, and seal. Store in a cool, dark place. This product will be ready for use in about two weeks. Pickled mussels and oysters turn dark if exposed to the light.

SHRIMP

Pickled shrimp is a regional specialty of the New Orleans area but is sold in fish markets from Key West to Washington, D.C. There are no standard recipes but those given here are believed to be typical.

Pickled or Spiced Shrimp Formula I

Peel the green shrimp and wash them well. Make a brine of 1 gallon water, 1/2 cup salt, 1 pint distilled vinegar, 1 tablespoon red peppers, 1/2 tablespoon cloves, 1/2 tablespoon allspice, 1/2 tablespoon mustard seed, and six bay leaves. Simmer the brine slowly for 1/2 hour, then bring it to the boiling point and add the shrimp. When they have cooked for 5 minutes, counting from the time the brine again begins to boil, they should be removed and allowed to cool. Pack in sterilized jars with a bay leaf, a few fresh spices, and a slice of lemon in each jar. Fill the jars with a solution made in the proportions of 2 pints of water, 1 pint of 5 percent distilled vinegar, and 1 tablespoon sugar. Seal the jars tightly and store in a cool, dark place. These pickled shrimp keep longer than those prepared by the second method, but the second pickled shrimp product requires less labor and is cheaper.

Pickled or Spiced Shrimp Formula II

Take 5 pounds of fresh, green headless shrimp. Wash them well but do not remove the shells. Put celery tops, salt, parsley, thyme, bay leaves, vinegar, and spices into the water. When this has boiled about 45 minutes, add the shrimp. Let boil 10 minutes, then set the pot aside and allow the shrimp to cool in the liquor. Drain and pack in small cartons with some of the spices. This product remains in good condition only a very short time unless held under refrigeration.

5 lb. shrimp	1 Tbs. allspice
1 gal. water	1 pt. vinegar, distilled
2 cups salt	1 Tbs. bay leaves
1 bunch celery tops	1 Tbs. red (chili) peppers
1 small bunch parsley	1 Tbs. black peppers
1/2 Tbs. cloves	1 blade mace

Chapter Sixteen

Caviar and Other Fish Roe Products

The roes or eggs of fish are among the most valuable of the miscellaneous food products from fishery sources. With few exceptions, almost any species may be utilized if the roe is large enough. The roes of sturgeon, salmon, mullet, alewife, whitefish, sea herring, cod, haddock, and tuna are the most important ones used in fish curing. A variety of curing methods may be used. Sea herring roe is air dried by Alaskan Indians. Sturgeon and salmon eggs are salted in brine and sold under the name caviar. Mullet roes are dry-salted, salted and air-dried, and brine-cured in the southern Atlantic states. Cod roe is salted and smoked in the northern European countries. Tuna roe is salted and air-dried in the Mediterranean area. Alewife and whitefish roes are cured as caviar in the United States.

CAVIAR

Caviar may be made from the roe of various species of fish but in the popular mind, caviar and sturgeon are synonymous. The reason for this belief is that sturgeon caviar was the first to be sold commercially, is the most important, and is the highest in quality and in price. This idea is so firmly fixed that the U.S. Food and Drug Administration has ruled that "it is believed that the name of the particular fish from whose eggs caviar is made from should appear on the label. In the case in point an appropriate label would be whitefish caviar."

At the end of the nineteenth century and in the early years of the twentieth, the United States had an annual production of 300,000 pounds of sturgeon caviar (Stevenson 1899). Most of this was exported to Germany (Stahmer 1925). Today less than 1000 pounds of sturgeon caviar are cured in the United States annually. Roe from other species is now cured in the United States and Alaska; for example, in 1940, 291,518 pounds of salmon caviar, 86,496 pounds of whitefish caviar (Fiedler 1945) and an undetermined amount from other species were cured. Instead of exporters we have become importers of caviar. In 1945, imports of caviar and other fish roe amounted to 86,337 pounds of sturgeon caviar, 144,463 pounds of other fish roe and 26,101 pounds of caviar, boiled (pasteurized and in hermetically sealed containers). (Anderson and Power 1946).

The American Caviar Industry

About fifty years ago the sturgeon was a common fish in American waters. In 1888 the catch of sturgeon was 17,879,000 pounds (Radcliffe 1925). In that year a single packer on the Delaware collected and shipped to Europe about 100,000 pounds of sturgeon caviar. When caviar and smoked sturgeon became popular, the prices of these products rose rapidly and fishermen found that sturgeon fishing was a very profitable occupation. As a result the waters were badly over-fished. In addition many small sturgeon, taken while fishing for other species, were destroyed. Some of the rivers, notably the Delaware and Hudson, once important for large catches of sturgeon, became so badly polluted that fish could not be there. As a result of all these causes, the sturgeon has been almost exterminated. Today the fisherman who catches a "cow" sturgeon usually prepares the caviar himself and sells it either to a local buyer or a dealer in the New York wholesale market.

In America, caviar is prepared from the roe of the following species of sturgeon: On the Atlantic coast, the common sturgeon (*Acipenser oxyrhynchus*); in the Great Lakes region, the lake sturgeon (*A. fulvescens*); on the gulf coast, the short-nosed sturgeon (*A. brevirostrus*); on the Pacific coast, the green (*A. acutirostrus*) or white (*A. transmontanus*) sturgeon; and in the Mississippi River valley, the shovel-nosed sturgeon

(*Scaphirhynchus platorhynchus*) are used. Roe from the last-named species is reported to make the poorest grade of caviar.

The sturgeon are caught in large mesh drift-gill-nets, similar to salmon gill-nets in construction and operation, and, especially in the Columbia and Mississippi river areas, on set-or trot-lines. A few sturgeon are taken in haul seines and traps. Gill net fishing is most effective at night and in localities, where the water is not clear. As the sturgeon are bottom feeders many of the sturgeon gill nets are operated as "sunken" gill nets, just scraping the bottom.

To make good caviar, the roe of the sturgeon should be taken just before it is fully developed or "ripe," and while it is still hard. If the roe is immature or "green" it does not make a good grade of caviar. Soft, fully developed roe in spawning condition also makes an inferior product.

The amount of roe obtained from an individual fish usually runs from 10 to 70 pounds, varying with size, species, and locality. Occasionally large specimens may yield more than 100 pounds of roe.

The "cow" sturgeon should be landed as soon as caught, the belly slit, and the roe removed. The roe must not be rinsed with fresh water as this tends to soften and break the shells of the individual eggs. It is also reported that roe washed in fresh water will not cure so well and is inferior in both texture and flavor. Exposure to direct sunlight is also a cause of deterioration. Rigid cleanliness and speed in handling are essential to a production of a good quality caviar.

The roe is placed on a work table with part of its surface made of a 1/4-inch wire-mesh screen. A second finer mesh screen below this slants at a 45° angle into a large tub. Some curers place a sieve with a 1/4-inch mesh directly over a large wooden mixing tub. The egg sac is slit carefully so that the individual eggs are not damaged and the roe is separated into portions. The curer rubs portions of the roe on the sieve gently, using the full palm of the hand. The individual eggs are separated from the membrane and fall through the mesh onto the screen below, sliding gradually into the tub. Slime, blood, and bits of membrane drain through the second screen. Vigorous rubbing is apt to crush many individual eggs, and to force pieces of tissue and other foreign material through the mesh of the sieve. The foreign material must then be separated from the eggs in the mixing tub, thus delaying the curing process.

Radcliffe (1925) suggests that:

> "Operators who prepare large amounts of caviar will find it convenient to fit the coarse sieve over a zinc-lined

> trough and rub the eggs into it. One type formerly used was 18 inches deep and 2 feet wide by 4 feet long. The bottom was sloping and had an opening at the lower end, closed with a sliding door underneath, the eggs being drawn off through the opening into the mixing tub."

The eggs, which look like small shot, are now ready for salting and mixing. Caviar makers claim that quality depends largely on skill in salting and the type of salt used. Some caviar makers with European experience claim that the best quality caviar can only be made if Lunenberg or Russian caviar salt is used. Others report excellent results with the mild-cure salt used for salmon. All agree that the salt must be distributed evenly over the eggs. This is accomplished by sifting the salt through a fine sieve, over the egg mass, using about 1 pound of Lunenberg or Russian salt or 1/2 pound mild-cure or packers fine salt to 12 1/2 pounds of roe. The New York caviar firms differ as to the amount of salt they believe is necessary. They give amounts ranging from 5 to 13 pounds of Lunenberg salt to each 100 pounds of roe. Users of American dairy or mild-cure salt are advised to add 4 pounds of salt only to 100 pounds of roe by one caviar merchant and are told that more than this will give a "woody" taste to the caviar. Another advises the use of the same amount of mild-cure salt as of Lunenberg salt. It is agreed that the cure should be mild as heavily cured roe will sell only to the cheapest trade, and that the amount of salt needed will vary according to the condition of the roe, the season (less salt is needed in cold weather), shipping conditions, and the species of sturgeon used. The amount of American dairy or mild-cure salt should be consistently somewhat less than the European salt since the former contains less chemical impurities, and therefore, penetrates more rapidly. European references commonly give 8 pounds of salt per 100 pounds of roe as an all around average (Hoffman (undated) and Jacobsen (1926).

As soon as the required amount of salt has been added, eggs and salt should be thoroughly mixed by hand. Paddles or other stirring utensils are not recommended. The process of stirring is the most delicate operation in curing caviar. Only trained men of much experience can do it really well, avoiding unnecessary breakage of eggs and securing an even blend of eggs and salt. At first the egg mass will be sticky but a brine is soon formed by the strong attraction of the salt for the liquid content of the eggs. Cobb (1919) stated that both hands should be used to mix the eggs and salt thoroughly for 5 to 8 minutes until foam or slime appeared on top of the egg mass. The mass is allowed to stand for about 10 minutes is then mixed again. By this time a copious brine should have formed, so that the eggs will pour readily. If the operation has been properly performed, a slight noise is perceptible when the mass is stirred, like small pieces of glass rubbing against one another.

The salted eggs are transferred to trays with a fine wire-mesh bottom (1/32-inch mesh) holding about 10 pounds each. These trays are placed between cleats on slanting boards set against a wall. They remain here until the brine is completely drained away. Sufficiency of drainage is determined by pressing against the underside of a tray at a single point. If the mass cracks open, the cure is completed and the mass may be removed. The time required for drainage is reported to vary from 2 to 4 hours. Good draining is very important, for it the caviar is "soupy" when received by the dealer it must be drained again, and the resulting loss in weight is charged against the shipper.

Radcliffe (1925) reports that:

> "The impression of some fishermen that the more salt added the heavier will be the caviar is wrong. The addition of salt extracts water from the eggs and reduces their weight. For the caviar to retain its delicious flavor the roe must be mildly cured. If an excess of salt has been used, the caviar on being treated by the canners will taste so strongly of salt as to be practically unsalable. The only changes in the amount of salt depend on temperature conditions. In cold weather as little as one pound of the Lunenberg salt or one-half pound of the dairy salt to 18 pounds of roe may be sufficient, but in very warm weather as high as one pound of the Lunenberg salt or one-half pound of the dairy salt to 9 pounds of roe may be required. If the roe is too ripe, and the entire egg mass is soft and tender to the touch, it may be put into a strong brine until it is thoroughly "struck" with salt. This, however, makes an inferior grade of caviar."

The sturgeon caviar is packed in small kegs to a net weight of 100 pounds. If the total amount to be shipped is less than this weight, wooden tubs or "kits," or large tin cans with tightly fitting covers may be used. Some curers line the containers with vegetable parchment paper before filling. The caviar should not be dipped out of the sieves in filling. Turn the sieve upside down over the shipping container and allow the contents of the sieve to fall out in a mass. The containers are not headed immediately but set in a cool place to stand for a few days until the caviar has settled. The head space thus formed is filled with caviar so that no air space is left, the top is fixed on tightly, and the container is ready for shipment. If possible, light and dark colored roe should be cured and packaged separately. Mixing the two colors gives the product a speckled appearance, making it less valuable than caviar of uniform color. Under no circumstances should any chemical preservative be used.

The caviar may be sold to wholesale dealers in the locality who in turn resell to packers and distributors of caviar in New York. Of recent years, however, some fishermen curers have been dealing directly with the New York packers. Caviar should be shipped and held in chill storage at 34° to 36° F. If the curer does not have refrigerated shipping facilities the container should be placed in large barrels or boxes and surrounded by ice.

When packaged for the United States retail trade, caviar is filled into cans holding 1 1/2 to 2, 4, and 8 ounces or into nappy glass containers holding about the same amount. Occasionally 1- and 2-pound containers are used, mostly for the hotel trade.

Caviar Industry of the U.S.S.R.

The U.S.S.R. has the most important caviar industry in the world and has devoted special study to the technology of caviar preparation and storage. The Caspian Sea is the only body of water in the world today where sturgeon are still to be found in large numbers. Fairly important amounts of sturgeon are also taken in the Black Sea, Sea of Azof, and in the rivers of the Ural region, according to Lazarevsky (Kasatkin 1940). The U.S.S.R. accounts for from 92 to 95 percent of the world sturgeon catch. Lazarevsky gives the U.S.S.R. production of various types of caviar in 1937 as: salmon caviar, 28,000 centners (5,236,000 lbs.) or 46 percent of the total; miscellaneous fish 18,800 centners (4,136,000 lbs.) or 38 percent; and sturgeon, 1,600 centners (352,000 lbs.) or 8.2 percent of the total. Although the amount of sturgeon caviar is the smallest of the different varieties, it is the most important in world trade, and its value exceeds that of salmon and miscellaneous types of caviar. It is also regarded as the highest in quality.

The caviar industry of southern Russia is based on 4 species of sturgeon. They are beluga, schip, ossiotr, and sevruga. The beluga (*Huso huso*) is the largest species and supplies the highest grade caviar. Beluga sturgeon usually weigh from 110 to 130 kilograms (242 to 286 lb.), and produce from 15 to 20 kilograms (33 to 44 lb.) of caviar. Occasionally a catch is made of a beluga weighing as much as 1,000

Qualitative Factors	Grades, finished product			
	Extra	*Highest*	*First*	*Second*
Degree of ripeness	Ripe	Ripe	Ripe	Not quite ripe roe permissible.
Egg membrane elasticity	Firm	Firm	Firm	Slightly weakened membranes permissible
Size of grain	*Large and med.*		*All sizes including small.*	
Color	Light gray, gray and dark gray.			Light gray,gray,dark gray and black.
Natural flavor	None	None	Very slight "grassy" flavor permissible.	Natural flavors permissible.

In all grades of sturgeon caviar, yellowish and brownish colors are allowed.

Table 1.—Raw material requirements in preparing sturgeon caviar (Kasatkin 1944).

kilograms (2,200 lb.) and containing as much as 160 kilograms (352 lb.) of caviar.

The schip (*Acipenser nudiventris*) and ossiotr (*A. guldenstadi*) sturgeon usually weigh from 15 to 80 kilograms (33 to 176 lb.) and produce from 2 to 12 kilograms (4.4 to 26.5 lb.) of caviar. The sevruga (*A. stellatus*) as the smallest of the sturgeons but is the most plentiful. It weighs from 12 to 50 kilograms (26.4 to 110.0 lb.) and produces from 1 to 5 kilograms (2.2 to 11.0 lb.) of caviar.

Sturgeon fishing may be carried on over most of the year with the exception of the legally established closed season, which is usually June and July. The season is at its peak in April and May, when the fish are just approaching spawning condition and come in to fresh water where they spawn among the reed covered flats of the river mouths.

The sturgeon roe is the most desirable for caviar at this time. There are two general types of sturgeon fisheries, salt and fresh water. In the shallower waters of the Caspian Sea a variety of set line or trot line is the most important type of fishing gear. These lines are often more than a kilometer (five-eighths mile) in length. The ground line is fastened to stakes driven into the bottom at each end. Hooks hang at the end of short lines fixed to the ground line at intervals. The hooks usually hang about 1 meter (3.3 ft.) below the surface, and are supported by cork floats. Gill nets are also used for the salt-water sturgeon fisheries and in some instances hand lines. In the fresh water sturgeon fisheries the haul seine is the most important type of gear, followed by the set line.

The fish are taken to a processing plant where the bellies are split open and the roe is removed. Great care must be taken to avoid contamination of the roe at this time. After removing, the roe is immediately rubbed through a "string" or metal sieve (fig. 26). This has a mesh large enough to permit the eggs to pass through without breaking but to retain tissue membranes. The eggs fall into metal pans below the screen.

The caviar is graded while it is raw, after being pressed through a sieve. It is graded first as to species, then as to the size of the eggs, their freshness, color, and flavor, their degree of ripeness, and the elasticity of the outer wall of the eggs. Grading is not exact and no two references agree. The most authoritative and recent is given by Lazarevsky (Kasatkin 1940). The data in table 1 taken from Lazarevsky,

Figure 26.—The Russian caviar industry, showing the butchering of sturgeon to remove roe and the separating of eggs from membrane.

showing the requirements of raw material for grain caviar in jars, indicates the grades and factors involved in grading the preparation of sturgeon caviar.

Unlike the German and American methods of preparing caviar, in the U.S.S.R. sturgeon eggs are washed before salting, except caviar that is prepared in strong brine, and packed in barrels. The eggs are washed in a pan with a perforated bottom through which water is forced. The water overflows, carrying off waste material. The loss in washing is usually from 1 to 3 percent of the original weight. When the surplus moisture is removed, the caviar is ready for salting.

The caviar is salted in pans. To obtain uniform distribution, salt is scattered over the caviar by shaking it through a sieve with 1 millimeter (0.059-inch) meshes. If chemical preservatives are to be used they are added at this time, mixed with the salt. Preservatives are added to all caviar packed in jars (except the grade "extra") as the preservative action of salt in concentrations from 3 to 5 percent is insufficient. According to the Institute of Fishery Technology in Moscow, the following mixtures of preservatives and salt have been used in the preservation o caviar.

Dosage of salt in percentage of raw material	Preservative	Dosage of preservative in percentage of raw material
4.5 to 5.0	Borax	0.5
	Borax and boric acid	0.5 (1:1)
	Borax and urotropin	0.6 (5:1)
	Sodium fluoride	0.1 - 0.2
	Sodium fluoride and urotropin	0.2 - 0.3 (1:1)
	Animalin [2]	0.2 - 0.5
	Solution of formaldehyde	0.2 - 0.3

Table 2.—Use of preservatives in Caviar.

The sturgeon eggs and salt are carefully mixed by hand. Sufficient salt is absorbed in from two to three minutes, depending on the quality of raw material, the temperature, and the quality and quantity of salt used. Sufficiency of salting is determined by the change in the appearance of the grains or individual eggs and the amount and consistency of the liquid formed. The excess liquid is separated from the caviar by pouring the contents of the pan over a sieve, which is shaken carefully. When properly salted the grains are globular and elastic because of swelling of the eggs and hardening of the membranes.

When salt is mixed with the eggs the moisture that is extracted dissolves the salt and forms a liquid. At the same time certain amounts of soluble protein are extracted. By the end of the process, when using a 4 percent concentration of salt, each 100 grams (3.57 oz.) of raw caviar, yields 4 cubic centimeters (0.135 oz. of liquid, which contains 13 percent salt and from 13.5 to 14.0 percent dry matter, the latter including from 0.5 to 1.0 percent of the substances extracted from caviar.

The quality of salt affects the processing as well as the flavor of caviar. The salt should be as free as possible from chemical and mechanical impurities. It should pass through a sieve with a mesh of one millimeter square without leaving a residue. Impurities, expressed in ions should not exceed: Cu. 0.05 percent; Mg. 0.02 percent; So. 0.40 percent, and Cl. content should not be less than 60.33 percent of dry salt. Insoluble residue should not exceed 0.05 percent and moisture should not be in excess of 0.5 percent.

In curing, caviar salting should be stopped when the maximum amount of moisture has been formed, but before any appreciable amount of soluble protein is extracted. This gives a product of maximum dryness with grains that are easily separated. Caviar made in this way is called "dry" or "grainy" as against "liquid" or "under-pressed" and "thick" or "over-processed." In the first case, salting was stopped too soon and not enough moisture was extracted from the Caviar. In the second case, the extraction of soluble protein went too far and the liquid was not sufficiently removed; it forms a thick mass which fills the spaces between the separate grains.

Dosage of salt without preservative (Percent)	Length of salting (Minutes)	Liq. formed per 100 gm. of caviar (Centimeters)	Salt in liquid (Percent)	Saltin caviar (Percent)	Substances extracted from caviar (Percent)
4.0	1.5	4.16	13.31	3.68	0.44
	2.5	3.47	12.08	3.83	1.52
	5.0	2.45	8.90	4.03	

Table 3.—Effect of salting time on amount of liquid and salt concentration (Kasatkin 1944)

The effect of the length of salting time on the amount of liquid and its salt concentration of extracted substances is shown in table 3. The salt content of the liquid is decreased, while that of the caviar is increased.

As soon as possible the caviar is removed from the sieves and packed in jars. The jars are slightly overfilled, the covers are pressed on, and the containers are set aside to allow additional free liquid to run off. The jars are set on top of the other not more than six high, and their positions are reversed periodically. The liquid stops running in approximately one hour. The covers are then forced down until they are in the proper place, care being taken not to leave an air space between the caviar and the lid of the container. The jars are next wiped with a towel and rubber bands placed around the tops between the cover and the neck of the jar in order to prevent penetration of air and moisture within the container. The sealed and cleaned jars are immediately taken to a refrigeration room for storage.

Grain Caviar in Barrels

Grain caviar in barrels is prepared when weather conditions do not permit packing of very lightly salted caviar in jars: when a more durable product is desired, as for export, or where refrigeration is lacking for shipment and storage. Grain caviar in barrels is also cured when the quality of raw material does not meet the requirements for caviar in jars. The quality requirements for grain caviar in barrels are less strict than for caviar in jars. There are three grades of the finished product, "highest," "first," and "second." Only three quality factors are recognized in grading. First, the degree of ripeness. This is the same for all three grades and simply requires that the eggs must not be over-ripe. The second factor is the size and color of the eggs. For the highest grade there should be little difference in size and color. For the "first" grade, eggs of the nearest colors or sizes can be mixed; that is, light gray with gray, large grain with medium. There are no size and color standards for "second" grade. The third factor is flavor. Muddy and grassy, or so-called natural flavors, are not permitted in any of the three grades.

The method of preparation of grain caviar in barrels is the same as for that in jars, except that the caviar in barrels is more heavily salted. The amount of salt used for this product varies from 7.5 to 10 percent of the weight of raw material depending on the location, season of the year, and quality of the product. Length of salting is from 5 to 8 minutes, the variation factors being the same as for the amount of salt.

The caviar is allowed to drain on wire-mesh screens for a longer time than that to be packed in jars. Surplus moisture must be more completely removed. Draining requires a period of from 2 to 4 hours. When sufficiently drained, it is packed in barrels lined with linen cloth or parchment paper. The barrels are set in chill storage for 24 to 48 hours to allow the caviar to settle, so that there are no air spaces. The barrels are then filled, headed, and shipped. While more durable than caviar in jars, it requires refrigeration at all times.

Pressed Caviar

The consumption of pressed caviar is confined largely to the domestic market, though some was exported to England before World War II. The Russian name for pressed caviar is "pa'yusnaya." Pressed caviar should be uniformly dark in color and soft in texture. The salt content should not be higher than 7 percent and moisture should not exceed 40 percent. The flavor should be pleasant, without bitterness or excessive sharpness, and there should be no "off" odor.

The first steps in making pressed caviar are identical with the preparation of grain caviar. The roe is rubbed through a metal screen to separate the individual eggs from the membrane. The eggs are then placed immediately in strong brine, without washing. Just before using, the brine is heated to a point where the temperature will vary between 40° and 42° C. (104.0° and 107.6° F.) after the eggs are added. The density of the brine should be 1.192 at 40° C. The quantity of brine should be four or five times that of the weight of the caviar to be salted. In order to prevent the brine from being weakened below the optimum point, a thin layer of large salt crystals is placed on the bottom of the tank.

The raw caviar is added to the concentrated brine and stirred gently. The length of the brining period depends on the quality and quantity of the raw material in a batch and the temperature of the brine solution but it should not exceed 2 1/2 and may be only 2 minutes. Sufficiently of salting is determined by observing the grains, which should be firm but not shrunken. When rubbed together they should grate like glass beads.

When the caviar is sufficiently salted it is scooped up in a sieve and packed in a long linen bag, which is then placed under a press. The degree of pressing is determined by the packer, depending on the size of the bag and the consistency desired in the finished product. When sufficiently pressed the caviar is removed from the press and is allowed to stand for 2 or 3 hours to permit cooling and uniform penetration of salt.

Since all operations should be carried out as quickly as possible, especially the removal of the caviar from the brine, the amount of raw caviar in any one batch should not exceed 25 kilograms (55 lb.). When the pressed caviar has stood long enough it is packed in shipping containers.

Pressed caviar is packed in oak barrels lined with linen and parchment, or in cans. Barrels used are of 50 kilogram (110 lb.) capacity and are reinforced with iron hoops. It is important to leave no air spaces in packing. The cans used are of the so-called sanitary type. They are lined with parchment and are hermetically sealed. The loss in pressing and packing pressed caviar averages 22 percent. Pressed caviar is somewhat more durable than grain caviar but should be kept under refrigeration

Pasteurized Grain Caviar

Caviar is pasteurized to obtain the maximum length of preservation. It is not considered as high in quality as fresh grain caviar and is intended for export, or use where refrigeration is not available. The original cure differs slightly from that for fresh grain caviar. The eggs are mixed with salt, then placed in a sieve as soon as the salt is dissolved, The brine formed by the salt in extracting liquid from the eggs is not poured off carefully as in packing fresh grain caviar and the caviar in brine is packed immediately into the containers. These are usually cans with a net capacity of 250 grams (9 oz.) and are inside enameled. Only dry, sterilized salt is used for curing the caviar.

A split-process pasteurization is given. The cans are processed for 90 minutes at 60° to 65° C. (140° to 149° F.). They are then cooled for 5 to 15 minutes, to 20° to 30° C. (68° to 86° F.) by a water spray and are then held for 24 hours at a temperature of 24° C. (73° F.). The pasteurization is repeated for a second and third time, following the same procedure. The cans are then washed, dried, labeled and packed in boxes. No chemical preservatives are used in the preparation of pasteurized caviar. Length of preservation is not indefinite, but the product will keep for several months at room temperature. If held at low temperatures, about 10° C. (50° F.) or less, it will remain in good condition for more than a year.

Cod Caviar

Cod roe is frequently made into caviar in the northern European countries to fill a demand for a lower-priced caviar. Cod roe is also occasionally made into caviar in the United States, especially for sale in the Boston, New York, and Philadelphia areas. This is done when other caviar is especially high-priced and scarce. There are several methods, of which the following is typical.

The fresh cod roe is freed from all bits of gall bags or viscera, discarding any dark or bloody roe. The roe is then mixed with 20 percent of its weight in salt and packed in barrels. The barrels are filled with an 80° salinometer brine, as the roe must be entirely covered at all times. If this is not done oxidation occurs during salting, giving the caviar an "off" flavor. The barrels are examined occasionally to determine the condition of the roe. When the salt has penetrated entirely through the roe, and it is completely cured, it is ready for manufacture into caviar.

The salted roe is drained and placed in new tubs where it is soaked in pure fresh water, preferably flowing, to remove the excess salt. If flowing water is not used, the water should be changed two or three times. From 6 to 10 hours are required for freshening. The roe is then put through a grinder after which it is mixed with water. The bits of membrane float to the top and are skimmed off. As an alternative, the lobes may be rubbed over a sieve, the roe passing through and the bits of membrane and other offal remaining.

Any bitter flavors may now be removed from the cod roe by placing it in a tub and covering with a sodium carbonate solution, made in the proportions of 98 liters (25.9 gal. U.S.) of water to 2.2 kilograms (4.8 lb.) of sodium carbonate. The roe is churned in this solution in which it is then allowed to stand for 15 hours. The roe is then filled into a bag which is hung up and left to drain. While it is in the bag the roe are washed several times in clear, fresh water. The roe should be sufficiently freed from sodium carbonate so that the drainage water does not taste alkaline.

The next step is coloring. To 1.5 kilograms (3.3 lb.) of pulverized gelatin which have been soaked in 80 liters (21.1 gal. U.S.) of water, are added 78 liters, (20.6 gal. U.S.) of hot water, stirring steadily until the gelatin is dissolved. When the temperature has dropped below 25° C. (77° F.), 1 kilogram (2.2 lb.) of sodium benzoate and 400 grams (14.3 oz.) of black dye are added. In some instances the benzoate is not added. Its use is not legal in the United States. The dye sometimes has a lamp-black base. Enough of this solution to completely cover the roe is poured over it and the roe is mixed thoroughly with the coloring solution. It is then allowed to stand in the coloring solution for 6 hours. At the end of this time the caviar is removed from the solution. It is understood that some packers mix lampblack or lampblack and caramel directly with the roe. This method requires much less time.

When it is colored, the caviar is aromatized or flavored with two solutions. The first is made by grating the yellow peel of lemons, which is then covered with 60 percent grain alcohol. After standing for about 10 days, when the flavor has been thoroughly extracted from the peel, the solution is filtered. Small amounts of the extract are mixed with caviar to taste. The second solution is made in the proportions of 500 grams (1.1 lb.) each of cloves, allspice, and coriander, with 250 grams (9 oz.) each of mace and ginger. These spices are pulverized and covered with 25 liters (6.6 gal.) of 30 percent alcohol. After macerating in the alcohol for several days, the spices are filtered out and the extract is added to the caviar in small amounts to taste, a few drops at a time. At this time some of the packers who do not use the dye bath method of coloring and sodium benzoate, in the proportion of 200 grams (7.14 oz.) of benzoate to 100 kilograms (220 lb.) of caviar. Boric acid is also used in addition to the sodium benzoate in the coloring solution by some packers. Urotropin (hexymethylaminetetramin) is reported to be used by some European packers. This preservative is prohibited under the U.S. Food, Drug and Cosmetic Act. The caviar is then packed in small vacuum-sealed glass tumblers holding from 2 to 8 ounces and in cans holding up to 1 or sometimes 2 pounds it is not heat-processed and should be kept under refrigeration for maximum preservation. Jacobsen (1926) gives a number of methods for making caviar from cod roe.

Cod Roe Salted and Smoked

Norwegian cod roe is obtained mostly in the Lofoten winter fisheries, where salted and smoked caviar is reported to be prepared as follows. The whole lobes of roe, fresh and in good condition, neither too ripe nor green, are cleaned of all slime, blood, bits of gall bag, and other offal. After being washed in sea water and drained for a few minutes, the roe are salted down in barrels. The roe must be salted whole without injuring or breaking the enclosing membrane, and must not be salted too much, just sufficient to keep them from decomposing and to give them an orange-red color. The amount of salt used is believed to average about 10 pounds per 100 pounds of roe. When the roes are cured, that is, when they have acquired the desired color, they are removed and washed thoroughly in sea water (although Stevenson says fresh water) several times. The roes are then hung out in the air on wires or drying racks, protected from the direct rays of the sun if it is bright. After drying about 24 hours, when no surface moisture is apparent and they feel dry to the touch, the roes are ready for smoking.

The roes are hung in the smokehouse on sticks, like fillets, or are laid on trays with wire-mesh bottoms, and cold-smoked for 2 or 3 days until they acquire a dark brown color. After smoking, the enveloping membrane or skin of each roe sac is split and removed from the lobe. The mass of separate eggs is then packed in barrels which are then tightly closed and placed in a normally cool place (about 55° to 60° F.), for a month or 6 weeks. At the end of that time fermentation should set in, which may be determined by the swelling of the barrel. The barrels should be watched closely during this period or fermentation may set in and progress too far before it is detected. As soon as it is determined that fermentation has begun, enough salt to arrest fermentation should be mixed through the roe. This is estimated at about 15 pounds per 100 pounds of roe.

Through the fermentation the roe is given a slight acid flavor and a taste resembling that of fermenting beer or wine, which must be stopped by adding salt at a certain point in the process, to be determined only through experience. The salt used to arrest fermentation must be fine and of the best quality. If the roe seems too dry, enough olive oil is added to moisten the product. After the roe has been thoroughly mixed with salt it is packed in tin containers, hermetically sealed but not processed. The usual size is a 1/4 round tin with a net weight of about 3 3/4 ounces. Glass jars are also used in the 1/4 size, with the same net weight.

Mullet Roe

From 80,000 to 150,000 pounds of dry-salted mullet roe are prepared commercially each year in the United States. This product is also cured in the Mediterranean area and in portions of the Soviet Union. Recent data on production in Europe are not available but it is believed that the pack of dry-salted mullet roe is larger than in the United States, where commercial preparation is divided between North Carolina and Florida. Cured mullet roe is a byproduct of mullet salting and fresh-mullet filleting in the south Atlantic and Gulf Coast states.

The quality requirements of the raw material are the same for the two methods of dry-salting followed here. That is, the roe must be fresh, of good color, and the skin of the roe sac must not be broken. The roe must not be over-ripe nor should it be too green or underdeveloped.

In both methods the first step is to prepare the roe by freeing the lobes from blood, gall bags, bits of intestine, and black skin. After being washed thoroughly and allowed to drain, they are ready for salting.

Dry Salting Mullet Roe

In the first method, after this preliminary treatment, the lobes of roe are rolled in fine salt. As a rule, 2 pounds of salt to 10 pounds of roe are sufficient. Too much salt must not be used as an excess will break the egg sacs. The lobes are picked up with as much salt as will cling and packed in tubs or boxes, with a scattering of salt over each layer. They remain in salt for at least 48 hours and then may be placed in chill storage and held in salt for some time if not marketed immediately. Sometimes salted-mullet roe is held in dry salt at room temperature. In packing for sale, it is taken out of salt and placed in fillet boxes with a light scattering of fresh salt. The consumer freshens the roe to suit the individual taste.

Air-Dried and Pressed Mullet Roe

In the second method the washed and drained roe are rolled in fine salt, using about 1 pound of salt to 10 pounds of roe. The lobes are removed from the salt in from 6 to 12 hours and are brushed well to remove any traces of excess salt. They are laid out to dry in direct sunlight, usually on a shed roof. During the first day of drying, the lobes of mullet roe are turned at least every hour and in the evening are brought indoors. Any night moisture such as dew, fog, or rain falling on the roe after drying has started will spoil or at least damage the product. Boards and weights are placed on the roes during the first night or two in order to compress them slightly. Curing requires about one week under good drying conditions. Drying is completed when the roe is reddish-brown in color and feels hard. The thumb should make no impression. The dried roe is usually dipped in melted beeswax which acts as a protective covering and is recommended as an additional aid in preservation.

After being allowed to cool for about 15 minutes the roe is wrapped in waxed paper, packed in wooden or tin boxes and stored in a cool, dry place. The product is sliced thin and eaten as an appetizer without further preparation.

This method of preparing roe is believed to have been introduced into the United States in colonial times by Englishmen who discovered it in Greece. Mullet roe, salted or dried according to this method, was well liked in England during the 17th century when it was called "bolargo." It is mentioned in Pepy's "Diary." Dried mullet roe is sometimes known as "botarque" in the Mediterranean area.

Salmon Caviar

Salmon caviar was originated about 1910 by a fisherman in the maritime Provinces of Siberia, and the preparation is a modification of the sturgeon caviar method (Cobb 1919). Salmon caviar has found a good market in the U.S.S.R. and other European countries, where it is known as "red caviar" to distinguish it from the sturgeon or "black caviar." Although several attempts have been made to manufacture salmon caviar in the United States, only a few firms in the Pacific northwest have operated successfully on a commercial scale. Their product is marketed mostly in New York and other eastern cities. A salmon-canning firm operating in the Bristol Bay area of Alaska also prepares salmon caviar, principally for export.

To be suitable for caviar, the salmon eggs must be absolutely fresh, free from blood, and of clear color and good consistency. Large eggs do not make good caviar. Most salmon caviar is prepared from the roe of silver and chum salmon, which have been found best suited for the purpose (Jarvis 1935).

The egg sac is split and rubbed gently over a table stand with a top of half-inch mesh screen. This mesh is just large enough to let the eggs drop through, separating them from the membrane. The eggs fall onto an inclined screen of fine-wire mesh leading to a large shallow box. The eggs drain on the screen and finally slide into the box. The eggs are cured in brine testing 90° salinometer, usually made from fine mild-cure salt. The salmon eggs are stirred occasionally with a wooden paddle to insure thorough mixing and equal absorption of brine. The brining time varies with season, temperature, and humidity, besides size, consistency, and freshness of the eggs. The time required varies from 15 to 30 minutes. The packers determine the sufficiency of the cure by noting the change in consistency of the eggs. The interior must coagulate to a certain jelly-like consistency but the eggs must not be shrunken. After brining, the eggs are dipped from the vat, placed on wire-meshed screens and drained overnight, or for a period of about 12 hours.

After draining, the eggs are filled into small kegs holding about 100 pounds and lined with vegetable parchment paper. The kegs are covered and allowed to stand until the eggs settle. The headspace caused by settling is then filled with more caviar, the kegs are headed, and put in chill storage at 34° to 36° F. until shipped. They are shipped under refrigeration.

The caviar is repacked in glass by large wholesale dealers in the eastern part of the United States. Nappy glass jars, holding 2 to 4 ounces are probably the most widely used containers. To obtain the maximum preservation the containers should be held at temperatures not higher than 40° F. or less than 29° F., which may keep the caviar in good condition for a year.

Salmon Caviar Russian Method

Chum and pink salmon are used most widely. Caviar may also be made from silver-salmon roe. Chinook salmon are not favored because of the large size of the eggs. The roe sacs are slit and rubbed over a screen to separate the eggs from the membranes. The eggs are then mixed in a concentrated salt brine (sp. gr. 1.200), previously boiled, and cooled to a temperature of from 13° to 18° C. (55.4° to 64.6° F.). The volume of brine should be three times that of caviar. Salt requirements are the same as for sturgeon caviar. The salting time varies as follows: In the Amur district, it is from 8 to 10 minutes for the best grade caviar; in the Kamchatka district, 12 to 14 minutes. For second grade caviar the time is 10 to 12 minutes in the Amur area, and 14 to 15 minutes in Kamchatka.

When sufficiently salted the caviar is allowed to stand for 12 hours to allow the brine to drain off and to permit uniform penetration of salt. Dry borax and urotropin are then added and distributed uniformly by mixing. Olive or cottonseed oil is added in small amounts and mixed with the caviar to prevent the grains from sticking and to give the product a more glossy, attractive appearance. The salmon caviar is then packed in barrels, which have been coated inside with a mixture of paraffin and wax, in equal parts. The sides and bottoms of the barrels are covered with parchment soaked in concentrated brine, then with cotton cloth impregnated with vegetable oil. Low-temperature storage is necessary for salmon caviar if it is to be preserved for any length of time.

Salmon Eggs for Bait

The first commercial use made of salmon roe was the preparation of bait eggs for catching trout and other game fish. This industry began about 45 years ago. A number of firms in the Columbia River and Puget Sound districts are now engaged in packing bait eggs and find a ready market for their product.

Each packer of bait eggs uses his own formula. The method described here is not used by all packers of bait eggs but has been used commercially and will serve at least as a basis for experimentation. The first steps in preparing bait eggs are almost identical with the first stages in the preparation of salmon caviar. The roe must be fresh and not too green; that is, the eggs must have a firm consistency and separate readily from the enclosing membrane. They must be of good size and clear in color. The egg sacs are split open and hot water is poured over the roe, which separates most of the eggs from the membrane and the remainder are freed by rubbing over a wire-mesh screen. This method is quicker than screening the entire mass but should not be used except by skilled bait-egg packers, as over exposure to hot water will damage the texture of the eggs and make them unfit for use.

Single Eggs

After draining for a few minutes, the eggs are placed in a solution of salt, sugar, and coloring material. The proportion of salt to sugar varies with different packers and may range from 1 part sugar to 3 of salt, to 1 part sugar to 9 of salt. The coloring material used is one of the aniline dyes in several shades of red, such as Sherwin Williams 3R or Erythrosin. The curing solution will test 80° to 90° salinometer and eggs will be left until cured. The cured eggs must have a firm consistency so they cannot be easily stripped from the hook but must not be shrunken, hard, or rubbery.

The color of the eggs depends on the strength of the dye and length of time the eggs are left in the dye-pickling brine. For a medium shade the eggs are left in the brine from 20 to 30 minutes. Some markets desire a brilliantly colored egg, while others require a pale shade resembling the natural color of the fresh egg. It has been the tendency among fishermen to insist that the first shade of color used for bait eggs in their district is the only one that will attract fish.

After the eggs are sufficiently colored and cured, they are drained on wire-mesh, screen-bottom trays and packed in small glass or tin containers. The most popular container is a glass jar with a two-piece screw top, holding 3 1/2 ounces. The tin can most widely used is a lithographed container of the same net weight. Sometimes a solution of glycerine and preservative is added before sealing the containers but packers of bait eggs claim that if the eggs have been properly prepared, this should not be necessary. Bait eggs will remain in good condition for a year if kept in a cool, dry place. If the jar is opened, however, mold will start growing in a short time (Jarvis 1935).

A preservative, usually one percent sodium benzoate, is added to the curing solution if the eggs are to be sold in warm climates. Also, the glycerine solution may contain 5 to 10 percent formalin for maximum preservation. The eggs will harden and turn white if the formalin solution is too strong. Formaldehyde is legally prohibited for use as a bait-egg preservative in some localities on the belief that it is poisonous to fish life. There are no scientific data to support this theory.

Cluster Eggs

In addition to single eggs, "cluster egg" are marketed. These are portions of roe with membrane and eggs adhering. They are used principally for fishing for steelhead trout. Cluster eggs are manufactured from small green roes, which are partially undeveloped. The clusters are washed, then placed in a brine as described in the preparation of single bait eggs. Preservatives, either formalin, hexymethylaminteramin (urotropin) or sodium benzoate are frequently added to this brine.

The time required for curing is variable and is determined by testing the consistency of the eggs. As a rule it is somewhat longer than that required for the preparation of single bait-eggs. When the pieces of roe are cured they are drained and packed dry in 1/2 or 1 pint glass jars. Under ordinary conditions of temperature and storage, "cluster" eggs will remain in good condition from 6 to 12 months.

Salted and Air-Dried Tuna Roe

Tuna roe is cured in southern Spain, in the Mediterranean area, especially on the Spanish coast, and to a lesser extent in French Africa. It is cured occasionally in our waters by foreign-born fishermen for home use but is not a commercial product in the United States.

Spent roe is not used for salting and drying. Only the roe of unspawned fish is desired. It should not be too small and unripe, nor be in the spawning stage. The roe is taken out of the tuna as soon as the fish are landed and should be prepared immediately. Great care must be taken in gutting so as not to damage the roe or taint it by breaking the gall bladder.

There is apparently some variation in the method of cure, according to locality (Dienzeide and Novella 1942; Classen 1946; and Uriarte (1926). According to one source the tuna roe is washed when removed from the fish and allowed to drain for a few minutes. It is then put into a tank partly filled with saturated salt brine (100° salinometer or 25° B.). Additional salt is thrown in so that the brine will not be weakened too much by extraction of moisture. The roe is left there about 12 hours. It is then removed from the brine, rinsed, and wiped off with the hands. Then the lower part of each lobe is pierced with a knitting needle to allow the moisture to escape. The roes are then placed in rows on a marble slab sprinkled with three-quarter ground salt. Additional salt is scattered over the roes and a second marble slab sets on top. After several hours, or, at the latest, the next morning, the roes are weighted down more heavily. Two days later the weights are removed, the roes are turned over, and the lower parts of the lobes are pierced again to allow the moisture in the tips to escape. After sprinkling with fine salt for the second time the roe is weighted down once more and allowed to stand 4 or 5 days.

The weights are then removed from the roes, which are rinsed quickly in a strong brine (almost saturated) and prepared for air-drying. Two parallel slivers of cane are placed at the narrowest part of the lobes, where they grow together, and are tied. A loop of string is fastened to this frame, with a hook at the end of the loop so that the roes can be hung from a line for air drying. After drying in the shade for several days, the time varying according to temperature, humidity, air circulation, and size of roe, the roes are hard and reddish-brown in color. When brushed with olive oil, or coated with beeswax, they are ready for market (Dieuzeide and Novella 1942).

In Spain the salting and drying of tuna roe is more of a large-scale industry than in France. As soon as the roes are separated from the viscera of the tuna, they are sorted and graded. Only the large and medium roes are taken for curing. Small roes are canned.

The first step in preparation is to clean the roe. The oviduct and the big lump of adipose tissue adhering to the roe are cut off. The large vein is removed and the blood from the small veins is forced out by pressure from the blunt back of a knife blade. The lower end of each lobe is punctured in several places so that moisture may run off during the drying process. After washing thoroughly in sea water the roes are allowed to drain a few minutes, then salted down in kenches on the concrete floor of the fish house. Each roe must be completely covered with salt. The kenches are from 2 to 3 feet high. he tuna roe is left in salt from 24 to 36 hours. It is then taken out of the salt, rinsed in sea water and allowed to drain, then placed in rows on a large board that has been sprinkled with fine salt. The roe is sprinkled with salt, a board sprinkled with salt is placed on top, repeating until there is a stack of six or seven layers. The completed stack is set in a simple screw press with large cross beams and placed under light pressure.

Classen (1946) reports the following process:

> "Every day the roes are taken out, rinsed with sea water, and the salt renewed (each day a lesser quantity of salt being used), and put again in the press with the pressure increased. After a week or so no salt is added and

> the pressure is again increased. After a few more days (in all, the roe remains in the presses 9 - 10 days) the roe is finally taken out from the press, washed, and rubbed very thoroughly in fresh water with a hard fiber brush to remove any impurities, salt, and slime, and hung on the beams of the drying ground in the same way as mojama, in the shade of rush mats. The color of the roe changes during the drying process from pinkish-yellow to a deep red-brown on the outside and a rich orange color inside."

An average of 15 days is required to dry the roe. It is then taken down and packed in paper-lined wooden boxes, ready for shipment. Sufficiency of drying is tested by pressing the lobe between thumb and forefinger. When the thumb makes no impression, the cure is finished. Classen estimates the average weight of the roes before salting as 4 kilograms (8.8 lb.) a pair, and that in curing and drying they lose about 60 percent of their weight, so that a pair will have a dried-weight approximately 1.6 kilograms (3.5 lb.).

Salted and dried tuna roe is regarded as an hors d'oeuvre of high quality. To serve, it is cut in thin slices and mixed with sliced onion and vinegar, or marinated in oil and vinegar.

Tuna Caviar

A paste made from small, deformed or damaged tuna roes is reported to be sold commercially in Spain under the name of tuna caviar (Classen 1946). The method of preparation is as follows:

The roes are washed, split open, then boiled in a brine made in these proportions:

80 liters (84.5 qt) water	1 kg. (2.2 lb.) sliced onions
2 kg. (4.4 lb.) salt	8 bunches celery
8 sliced peeled lemons	10-12 large green Spanish peppers

Black peppers, garlic, bay leaves, to taste

The liquid is brought to a boil. The roes, in a wire basket, are immersed in the liquid. The pieces stick together very easily, causing uneven cooking, and therefore, must be stirred gently from time to time. If stirring is not done carefully many grains separate from the mass and are lost. When thoroughly cooked, about 20 minutes of simmering, the baskets are taken out and set aside to drain and cool. When the pieces are cool enough to handle, the eggs can be separated from the fibrous tissue in lumps without the use of a knife. The next step is to pass the eggs through a meat grinder, using the plate with smallest holes. The eggs now have a pinkish-yellow color. They must be handled immediately as they turn gray if left exposed to the air. The ground roe and other ingredients are placed in a mixing machine together with the following:

30 kg. (66 lb. ground roe	300 gms. ground sweet paprika
2500 gms. (5.6 lbs.) lard	1 qt. vinegar (2 1/2 %)
300 gms. (11 oz.) salt	15 gms. (0.5 oz. ground cloves
45 gms. (1.5 oz.) ground white pepper	2500 gms. (5.6 lb.) ground onions, fried golden brown in lard or oil.
3 liters (3.2 qts.)stock used in cooking roe, filtered	
3600 gms. (8 lbs.) canned sweet peppers, ground	

The ingredients are thoroughly mixed until a paste of satisfactory texture is obtained. The paste is filled immediately into size 1/8 cans with a capacity of 100 grams (3 1/2 oz.) net. The cans must be well filled, so that no air remains in the can, but care must be taken to avoid overfilling. The cans are sealed and heat processed. The process is 108° C. (266.4° F.) for 50 minutes.

Miscellaneous Caviar

Caviar is made from the roe of a number of fish besides those previously mentioned. In some parts of Europe the roe of carp, herring, perch-pike and other miscellaneous fish are made into caviar. In the United States some river herring or alewife roe is manufactured commercially as caviar. In fact, the roe of nearly all sea fish and many of the fresh-water fish is used if it is obtained in a suitable condition. Nearly every curer has his own method and the preparation differs somewhat with different species. As the individual eggs of nearly all species are small there is no accurate gauge as to sufficiency of salting. Because the products must be rather heavily salted for long preservation, many of them are much too salty for the average American taste. A more careful control of salting would do much to improve the popularity of caviar in the United States market and promote the utilization of miscellaneous fish roe.

One of the European methods used in making caviar from the roe of herring, pike, and other miscellaneous fish is given below. If followed carefully the method may also be used with alewife roe in the United States. The fresh roe is cleaned carefully and freed of blood and slime. It is then rolled in fine salt and packed in barrels which are filled with strong salt brine and headed up. The roe is held in salt until it is to be made into caviar. The salted roe is then washed thoroughly in cold water to remove surplus salt and other extraneous material. A solution is made up in the proportions of 5 liters (5.3 qt.) of cold water in which is dissolved 50 grams (2.0 oz.) of borax to 10 pounds of roe. The herring roe is soaked in this solution for 24 hours. It may be changed once or twice during this time.

When the freshened roe has drained, the skin of the roe sac is split and the eggs are rubbed over a sieve with a 1/8-inch mesh. The eggs pass through while the skin and connective tissue are retained. The roe is again washed thoroughly then drained until it is free from water. The roe is then colored. The coloring material is lampblack or lampblack and caramel. A mixture is made up in the proportion of 50 grams (2.0 oz.) of salt, 10 grams (3/8 oz.) of pulverized citric acid, and 10 drops of oil of lemon, to each 10 pounds of roe. The flavor is also improved if 100 grams (4.0 oz.) of olive oil is added. The mixture is stirred thoroughly into the roe. After standing 6 days, the caviar is ready for use. It is packed in vacuum-sealed glass tumblers with a net weight of two and four ounces. The jars are not heat processed.

Chapter Seventeen

Smoking Fish

Smoked fish has served as a staple food since the dawn of time. If an explanation of its origin were called for, it would have to be some legend such as Charles Lamb's story of the discovery of roast pig. In all probability the discovery of the use of smoke as a method of preserving flesh foods such as meat and fish was accidental. It was certainly discovered long before man had progressed beyond the Stone Age of existence. The logical deduction is that smoked fishery products are a natural development of drying. Smoke from the fire in the hut flavored dried fish hung from the rafters for storage. When the improvement in flavor and length of preservation were noted, smoking as a method of food preservation was developed. The commonly accepted story of the discovery of "finnan haddies" lends support to such a theory.

Though the quality of smoked fish used, and the value, have increased, smoked fish has lost in relative importance to other fishery preparations as a food product in this country. The increase in value is due to the development of smoked salmon and a few other high-priced specialty products. The consumption of the lower priced standard products, most of them as appetizing as smoked salmon, has not increased appreciably. At one time such an article as hard-smoked or "red" herring was a staple article of food in the interior as well as along the seacoast. Now, smoked fish is bought as a regular trade article of diet only by some of the inhabitants of certain seacoast areas, or by immigrants who have not yet lost their taste for this product.

With better advertising and merchandising methods, backed by a carefully prepared product of standard quality, the smoked-fish trade of this country would be considerably increased. A few firms which are making efforts in this direction are marketing several types of smoked fish, formerly sold in bulk, in an attractively packaged form. The reception of such articles as selected bloaters, trimmed of head and tail and packed in cellophane wrappers, sliced smoked salmon packed with olive oil in quarter-oil sardine cans, in 4-ounce jars, and in packages of 1 to 8 ounces wrapped in cellophane, and choice kippered herring in cellophane envelopes is encouraging and indicates possibilities for building up a good market for quality smoked fish.

The development of smoked fish as a packaged article should include a more extensive use of the glass container as well as cellophane or parchment paper. Glass has been used to some extent since the early nineteenth century but a systematic study on the wider application of glass jars as marketing containers for different varieties of smoked fish, accompanied by the proper merchandising, would undoubtedly increase the sale of smoked fish.

CHEMISTRY OF FISH SMOKING

Several studies have been made in recent years on the chemistry of wood smoke in curing fish. These studies are not yet complete and there is some disagreement as to results. On the basis of present knowledge, preservation in smoking depends principally on the removal of moisture by drying and salting or brining, rather than the inhibitive action of different volatile chemical substances given off in wood smoke. A number of investigators have published data indicating that wood smoke has some bactericidal effect. It would seem, however, that wood smoke is more valuable as a flavoring than a bactericidal agent. When it is desired to prepare smoked fish with better keeping qualities than the average, the salting and smoking are done very thoroughly. Rector (1925) believes that with fish the preliminary salt treatment is even more important since it helps in the preservation and at the same time makes the fish firmer by drawing out large quantities of water.

The preservation action of wood smoke has generally been held to be largely due to the presence of a number of phenolic compounds classified under the wood-creosote group. In regard to this, Tressler

(1923) states that the wood-creosote constituents as determined in the products of wood distillation are the xylenols, guaiacol, creosol, and other methyl esters of higher phenols. There is also, possibly, a very small amount of lower phenols such as creosol and phenol, but they are thought to be present in very small quantities. The number of organic compounds occurring in wood smoke are very great and there may be other compounds deposited on the fish which also have a powerful preservative action. Some of the other organic compounds produced in wood smoke are formaldehyde, acetic acid, acetone, and methyl (wood) alcohol.

An important and interesting contribution to the chemistry of fish smoking has been made by Hess (1929) in a study of the bactericidal effect of smoke in fish curing. He believes that formaldehyde is an important disinfecting constituent of smoke. He finds little difference in the bactericidal effect of smoke from different sources, such as hardwood sawdust, softwood sawdust, or a mixture of both. Dense wood smoke was found to contain a higher concentration of preservative or disinfecting agents than a light smoke. Griffiths and Lemon (1934), in a study on the smoking of haddock, included a determination of the effect of formaldehyde in the preservative action of wood smoke. As a result of their experiments they came to the conclusion that though formaldehyde is found to be present in small quantities in wood smoke and smoked fish, and in the destructive distillation of sawdust, it has no significant preservative action in cold-smoking fish.

Liquid Smoke and Other Chemical Preparations

The use of chemical agents such as liquid smoke or of salt impregnated with chemicals found in wood smoke has been advocated as a substitute for smoking. It is claimed that smoke flavor is better controlled, flavor is improved, greater length of preservation is secured, the time required for curing is greatly reduced, and that the smokehouse may be eliminated. These agents have been found useful in meat curing, where they seem satisfactory, especially for non-commercial use. Experimental packs of fish using liquid smoke and two varieties of "smoked salt," that is, fish salt impregnated with the products of combustion of some one of the woods used in the smoking of fish or meat, were made at the laboratories of the College of Fisheries, University of Washington, and the U. S. Fish and Wildlife Service. The conclusion was reached that the appearance, texture, and flavor were not equal to that of fish smoked by common methods. It was also believed that while liquid smoke and smoked salt might be used as a flavoring in some types of canned fish or in the preparation of fish fillets, they are not a satisfactory substitute for the smokehouse.

Chemical Preservatives for Smoked Fish

It must be remembered that preservation by smoke-curing is temporary, the fish remaining in good condition for only a comparatively short period. Smoked fish is subject to mold or bacterial spoilage within a short time unless it is held at a low temperature. Attempts to inhibit spoilage by the addition of chemical preservatives have been made with more or less success. The preservative is sometimes added to the brine used in salting the fish preparatory to smoking. At other times preservatives are dusted over the smoked fish. The use of these substances is not advisable. Original inferior quality is too easily masked by the use of chemical preservatives. Then too, regulations governing the use of chemical preservatives in food are becoming more stringent as knowledge of their effect has accumulated. Sodium benzoate is one of the few preservatives permissible under United States law. It may be used in amounts not greater than 1/10 percent and its presence and amount must be stated on the label, with directions for its removal. Other preservatives that have been used in the preparation of smoked fish are mixtures of boric acid, salicylic acid, and salt. The use of these preservatives is now prohibited under the United States Food, Drug and Cosmetic Act.

General Methods

There are two general methods of smoking fish: cold-smoking or light-smoking, and hot-smoking or barbecuing. The products obtained by these two methods are different. Neither can be recommended over the other. Cold smoking is the most important method in the United States, Canada, the British Isles, and Norway. Hot smoking is used most widely in Denmark, Germany, and Sweden.

In cold-smoking, the fish are hung at some distance from a low smouldering fire and are smoked at a temperature usually lower than 80° F. In curing a few products the temperature may be as high as 90° F. but never more than 100° F. The degree of preservation depends on the length of time the fish are smoked. Fish cold-smoked a few hours, for example, will keep only a short time, not much longer than fresh fish; finnan haddie and kippered herring are examples. If an extended period of preservation is desired, the fish must be smoked for several days up to 2 or 3 weeks. Hard-smoked or red herring is an example of cold-smoked fish which may be held for a long time. In hot smoking or barbecuing the fish are only a short distance from the fire, and are wholly or partially cooked. The temperature in hot smoking will vary from 150° to 250° F. The cure is completed in a few hours, usually not more than 3 or 4, and sometimes only 2 hours. Hot smoking arrests deterioration temporarily since the enzymes causing autolysis of the flesh are destroyed by the heat of cooking. Hot-smoked fish is ready for consumption when cured, requiring no further cooking. In the United States, hot smoking is used less than cold smoking. It is used principally for preparing kippered salmon and sablefish on the Pacific coast, smoked whitefish in the Great Lakes area, and for smoked eels and sturgeon in the North Atlantic coastal area.

Dressing or Preparation for Smoking

Fish may be smoked either round, gibbed (gills and part of the viscera removed through a small opening at the throat), gutted, split, in fillets, or cut into small pieces. The skin may or may not be removed, depending on the product and on individual preference. Small fish are usually smoked round. Alewives and herring, except for kippers, and some lots of large herring, which are gibbed, are usually not dressed at all. Lake herring, eels, whitefish, and flounders are usually cut down the belly to the vent and gutted, but the head is retained. Haddock and mackerel (both Boston and Spanish) are split so that they will open in one piece. Cod and some large haddock are cut into fillets. Salmon are split into sides. Angler-fish, carp, halibut, and sturgeon are cut into small pieces for smoking. Sablefish and salmon are cut into small pieces for kippering

Salting

After butchering, fresh fish are washed, drained, and salted for varying lengths of time. Salting is an essential and important feature of smoking, as smoking is not a sufficient preservative in itself. Unsalted fish will usually sour or spoil under temperature and humidity conditions in the smokehouse before they can be cured. Then, too, the salted product is much more appetizing. Even if temperature and humidity conditions are favorable, a great deal more time is required for curing if the fish are not salted. Salt removes moisture much more rapidly than by drying in smoke. The length of time in salt depends on the species of fish, flavor desired, length of preservation required, and trade preference. Salting may be done either by immersion in brine or by scattering dry salt on the fish. It is not necessary that the fish be struck through as when they are to be entirely salt cured. A few hours are sufficient, depending on the climate and product desired. Some fishery products, such as smoked fillets, are salted only from 15 to 30 minutes. Fish such as halibut, herring, mackerel, or salmon may be smoked after being held in salt for a year. In these cases the excess salt is removed by soaking the fish in fresh water, though sea water is sometimes used. Fish given a heavy salt-cure and held in storage for some time before smoking are not so desirable in quality as those given a light salting and smoked immediately. In the Orient, fish are sometimes smoked without being salted. These fish are dried to a low moisture content and are not particularly appetizing to the occidental palate.

Drying

When the brining or salting is completed, the fish are washed or rinsed in fresh water to remove any excess salt. They are then hung on sticks, placed in trays, or arranged on movable racks which can be run into the smokehouse. Halibut or similar species may be dried for a day or two in the open air before being fastened to smoke sticks and hung in the smokehouse. It will be found that fish will have a more attractive appearance and require a shorter time in the smokehouse if dried in the open air for a few hours beforehand. In damp weather fish are sometimes dried by hanging in the smokehouse over a low clear fire with little smoke. It has been found that the use of electric fans or blowers is better than drying in the smokehouse in unfavorable drying weather.

The Smokehouse

There is no standard model for a smokehouse. Various types have been developed in different parts of the world. Many of the foreign types are in use in the United States. Some New York fish curers do not have a regular smokehouse. They smoke fish in warehouse rooms having brick walls, concrete floors, windows blocked, and the fuel burning in a metal trough on one side of the room; the smoke escapes through a skylight. Studies made to develop improved equipment for fishery products are limited. Lemon (1932) devised a smokehouse in which humidity and temperature could be controlled. Workers at the Atlantic Coast Experimental Station, Halifax, Nova Scotia, have developed a smokehouse which is expected to overcome many smoking problems and result in an improved product (Cooper 1937; Linton and Wood 1943). It is now being used commercially in the Maritime Provin-

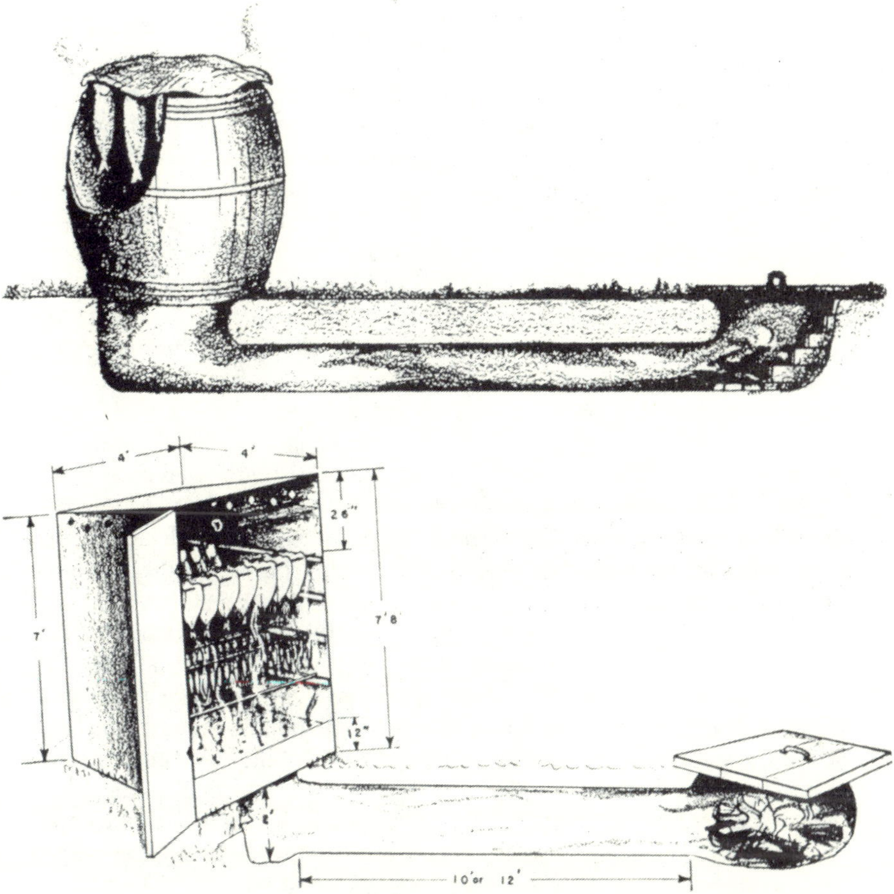

Figure 27.—Small smokehouses: Barrel smokehouse; shed smokehouse.

ces, Canada. After a study of the disadvantages of old-type smokehouses and of the variables requiring control, a smokehouse based on aerodynamic principles was invented at the Torry Research Station, Aberdeen, Scotland, (Cutting 1942; and Anon, 1946b).

The product to be smoked, the size of the market and shipping area, often determine the style and size of the smokehouse. In smoking some articles, such as finnan haddie, the smokehouse has special features which will be described under the discussion of the individual products. The smokehouse may be a small and simple structure when fish is smoked for local use, to obtain experimental data, or for introducing smoked fish to a new market.

A smokehouse for preparing small lots of fish is easily made by knocking the ends out of a hogshead or large barrel, setting it over a hole in the ground about 2 feet deep and a little narrower in width than the diameter of the barrel. Wooden strips are nailed inside the barrel on two sides, a few inches below the top. The ends of the smoke sticks rest on these strips. A loosely fitting wooden cover is placed on top, or it may be covered with damp sacking. Approximately 12 inches of the lower ends of a couple of staves may be removed to furnish a door to the fire pit, or a hole may be dug adjacent to the fire pit and covered with a cover. The fire is fed through this hole, which also serves as a draft when the lid is partly raised. A smokehouse of this type is best for hot smoking but may be used for cold smoking if operated carefully. If the fire is permitted to flare up, however, the fish are soon scorched. For cold smoking it is best to dig the fire pit about 12 feet away from the smokehouse pit, with a covered trench between, which acts as a pipe. The fire pit should be placed on the side from which the prevailing winds come (fig. 27).

If a more permanent smokehouse is desired, and one that will handle a larger amount of fish or about 200 pounds at a time, a wooden shed is made 7 feet high and 4 feet square, inside measurements. Resinous wood should not be used in construction. About 12 inches above the ground a false bottom is placed with 3/4-inch or 1-inch auger holes spaced at 2-inch intervals. On the two sides wooden battens are nailed spaced 12 inches apart, the first about 18 inches below the top. The ends of the smokesticks on which the fish are hung, or the wire-mesh bottom trays, if they are used, rest on these battens. The whole front of the house is hinged for a door. Three or four holes about 2 inches square are cut on the two sides a few inches below the roof, sometimes with slides to cover for use as drafts or ventilators. This shed is set over a pit about 2 feet deep, connected with a fire pit by a 10-foot covered trench. The pit below the smoke-house may be lined with brick. A terra cotta drain pipe, to act as a chimney, may be placed in the trench connecting the two pits (fig. 27).

The smokehouses used for curing herring in Maine and the adjoining Canadian Province of New Brunswick are barnlike wooden structures, with a louvre running the length of the ridgepole. The interior is divided by a series of vertical rows of two-by-fours. The spaces between the rows are called "bays." The scantlings in each row begin about 2 feet below the ridgepole and extend horizontally crosswise of the building, with the rows spaced from 12 to 18 inches apart. They are placed within 6 to 8 feet from the ground. There is no floor as the fires are built on the ground. The average size herring smokehouse is about 30 feet long, 18 feet wide, and about 25 to 30 feet high.

Some of the smokehouses used in New York, especially for hot smoking, are like the "ovens" used in Germany. They are made of iron sheeting about 9 feet high and from 4 to 6 feet square. The doors cover the whole front. Doors are usually of the two leaf type and may be divided in two more sections like a Dutch door. Drafts are usually

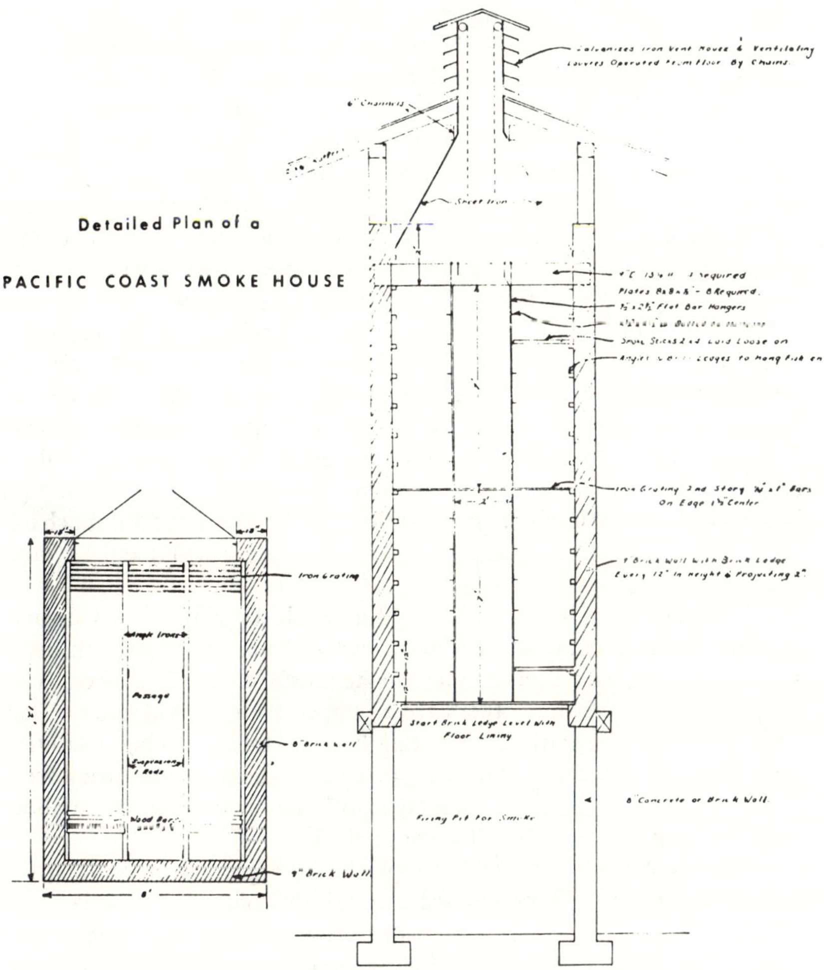

Figure 28. – Plan of Pacific coast commercial smokehouse.

built in the lowest section. Several ovens are built in a row and are sometimes held in a framework of brick. In some cases, the principal fuel is gas supplied by perforated pipes just above the floor on the two sides. Metal plates are fixed a few inches above the pipes. The plates are covered with sawdust which is charred by the gas flame below, when it is necessary to produce smoke. The gas flame is used for drying or cooking. In other instances gas is not used. Wood chip and sawdust fires are built inside the house.

The smokehouses used in the Middle West for curing such fish as cisco, lake herring, tullibees, chub, whitefish, trout, and sturgeon are much like those described in the preceding paragraph, and usually consist of 3 or 4 smoking chambers built side by side. These chambers are constructed of brick with 8 inch walls and a ceiling of tinplate or galvanized sheet metal. Their dimensions range from 6 to 14 feet high, 3 1/2 to 5 feet wide, and 6 to 12 feet long (inside measurements). Most of these smokehouses are intended to smoke only 3 rows of fish, the lowest hanging from 3 1/2 to 6 1/2 feet above the floor. The other rows are placed at intervals of 13 to 18 inches (all variations in spacing given depend on the species smoked) above the bottom row. The top row should hang not less than 18 inches below the ceiling. If the space between the ceiling and the top row of fish is insufficient, the hot air collecting at the top of the smokehouse will cook or blister this row of fish, often softening them so that they will fall. A number of small holes, an inch or two in diameter, are made in the ceiling to allow surplus smoke to escape to the chimney. The fire may be built in a pit below the level of the floor or the ovens may be gas fired, fed mechanically with sufficient sawdust to furnish smoke (Tariff Comm., U.S., 1927).

Portable smokehouses developed and sold by middle west dealers in meat packers' supplies, for use in curing hams or bacon, are being used by some fish curers. The smokehouses are of the oven type. Some are gas fired while others are equipped with a wood grate. They have mechanical controls, to obtain any desired smoking condition regardless of weather. The price of one of these ovens varies from about $300 for the wood-burning type, with minimum controls, to about $500 for the gas-fired type, fully equipped with controls.

The smokehouse (fig. 28) is typical of the general style along the Pacific Coast in the United States and British Columbia, Canada. It is suitable for smoking a wide variety of fishery products on a commercial scale. This house is of brick and consists of a single chamber, with inside dimensions 12 feet long by 8 feet wide. The smokehouse is divided into two stories and a fire pit. The fire pit is 8 feet high and is separated from the first story of the smokehouse by an iron grating of 1/4 by 1-inch bars. The fire pit door is made of sheet iron and is built in two horizontal sections, with a draft to regulate the fire built in the lower section. The walls of the smokehouse are 8 inches thick with a brick ledge every 12 inches. A two-foot passageway runs through the center of the smokehouse. Four 2 1/2 by 1/2-inch flat bar hangers are fixed from the beams at the top of the smokehouse to the first floor of the grating, on each side of the passage. Angle irons (2 1/2 by 2 1/2 inches) are bolted to these hangers horizontally at intervals of 12 inches at the level of each brick ledge. The smoke sticks or trays are laid from angle iron to brick ledge, in loading the chamber for smoking. Ventilation is obtained through a galvanized iron vent house in the roof, furnished with movable louvres operated from the floor by chains. A recording chart thermometer, fixed to a side wall, provides means of keeping an accurate record of smokehouse temperatures, which is of value in obtaining data to determine processes.

The smokehouses of large commercial curers are about the same size, but from 6 to 10 instead of a single chamber will be built in a row. In some cases, the houses are equipped with movable racks which can be loaded or unloaded outside. This arrangement is more convenient and requires less labor for operation, but the capacity is reduced.

The smokehouse designed by Lemon (1932) is of the cabinet type, of fireproof material and is portable. The laboratory model built at the bureau of fisheries (now Fish and Wildlife Service) technological laboratory is 7 1/2 feet high by 3 feet square. The cabinet is divided into two sections, an upper enclosed compartment in which the fish are smoked and a lower compartment containing the operating mechanism. Smoke is produced by mechanically feeding sawdust on a metal belt drawn over an electric hot plate. The density of smoke is controlled by regulating the flow of sawdust. The smoke enters the upper section through the smoke pipe. A perforated false bottom or baffle and an agitator insure an even distribution of smoke. Two sets of thermostatically controlled electric heaters are attached to the upper side of the baffle and provide the necessary heat in the cabinet. A recording psychrometer makes possible the determination of the relative humidity, which can be controlled at any point between 8 percent and saturation by wetting filter cloths hung on the sides of the chamber.

The Atlantic Fisheries Experimental Station of Canada smokehouse is a tunnel-type structure, constructed mostly of wood, and is 14 feet long, 8 feet 7 1/2 inches wide, and 5 feet 3 3/4 inches high. A schematic drawing showing the principle of operation is given in figure 29. It holds two movable racks with wire-mesh trays on which the brined fish

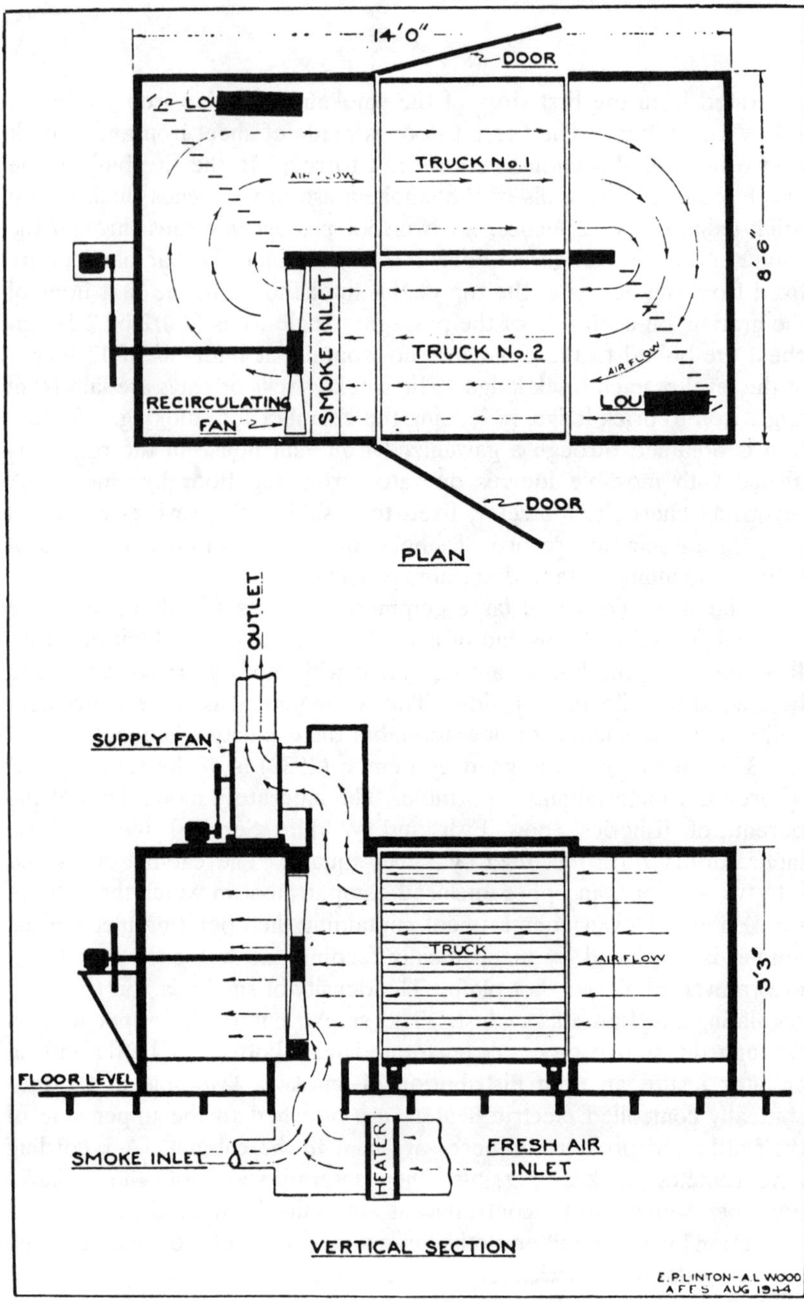

Figure 29. – Schematic drawing of smokehouse designed by the Halifax Fisheries Laboratory.

are placed. The individual racks are 3 1/2 by 4 by 3 1/2 feet, with a capacity of 400 to 500 pounds of fillets. This smokehouse may also be used as a salt fish drier. A number of these houses have been installed by commercial fish curers in the Maritime Provinces of Canada, with good results.

Linton and Wood (1945) report as follows:

> "Smoke is produced from sawdust and shavings in a simple type of cone smoker placed outside the house as shown in the sketch. An automatic type of smoker (Smokemaster) using sawdust is an alternative source of smoke. The supply fan draws air through the steam heater and smoke from the smoker into the house and expels an equal quantity of the waste air-smoke mixture. A propeller type fan, 42 inches in diameter recirculates the air and smoke over the fish within the house. With two cars of fish in the house the propeller fan blows the air-smoke mixture over one car of fish, through guide vanes and then over the second car of fish. The guide vanes at each end of the house are adjusted to direct the air and smoke evenly over the face of the cars of fish. The house was designed to offer little resistance to air flow and approximately one-horsepower is required for the operation of the two fans. The air pressure inside the house is always slightly less than the atmospheric pressure and hence the smoke does not escape readily into the surrounding room. The temperature of the house is maintained constant by controlling the heat added to the incoming air in the steam heater. The maximum steam load is about 5 boiler horsepower."

Ice or mechanical refrigeration is used to cool incoming air in warm weather. This is necessary to prevent overheating in the house and also to dehumidify the air. In smoking and drying the fish by this method, the weight loss is largely controlled by adjusting the volume of fresh air admitted, while the depth of color depends on the amount of smoke forced into the house. Thus, weight loss and color are made independent factors and any degree of dryness may be combined with any desired shade of color without the use of artificial colors. The time of smoking fillets has been reduced to an average of three hours. Anderson and Pederson (1947) have developed a new type of smokehouse, more efficient in operation than the old kiln type. Fish are smoked which has been adopted by about a dozen commercial fish curers in the Pacific northwest. This smokehouse is also of the tunnel type. Special emphasis is placed on simplicity of design and accurate control of the main variables in smoking: Smoke quality and volume, temperature, humidity, air velocity, and distribution. The smoke generator, usually a metal drum, is located outside the smokehouse, so constructed that the smoke is free from grit, soot, and ash. Smoke is delivered by a large diameter pipe, to permit condensation of tars and resins. The pipe line is also equipped with traps for catching and draining these materials.

The smoke enters a mixing chamber where it is mixed with air. The air-smoke mixture is drawn through an opening across heaters by a blower. The smoke is then distributed through the smoking chamber by means of the volume blower and baffle arrangement. After it has been deflected by the baffles, the smoke may either be exhausted through an outlet damper, or recirculated through the blower and heater. Humidity is controlled by installing a cooling system just ahead of the air intake.

The construction of the smokehouse, method of operation, and new processes developed for operation under controlled conditions are described in *Technological Report No. 1, The Smoke-Curing of Fish and the Application of a Controlled Method to the Process*, State of Washington, Department of Fisheries. This publication should be consulted for further information on the process.

Fuels Used

Hardwoods are generally favored as fuel in smoking, both for drying and the production of smoke. The pitch or resins contained in softwoods, such as pine, are volatilized in burning and collect on the fish, giving it a strong characteristic flavor which ruins it for most markets. In New York, however, fish given a short hot smoke may be cured over a fire of excelsior and softwood sawdust. Different varieties of wood are used as fuel for smoking fish in the various regions of the United States. Alder is a favorite fuel on the north Pacific coast, oak or manzanita roots in California. White birch is preferred for smoking herring in Maine, although old driftwood, a mixture of all sorts of wood brought in by the tide, may be used. Some of the woods that may be used for fuel for smoking fish in the southern states are scrub oak, live oak, hickory, river mangrove, and palmetto roots. Oak and hickory are favored throughout the Middle West. Corn cobs have been found to make good fuel for smoking fish in this region but the fire must be

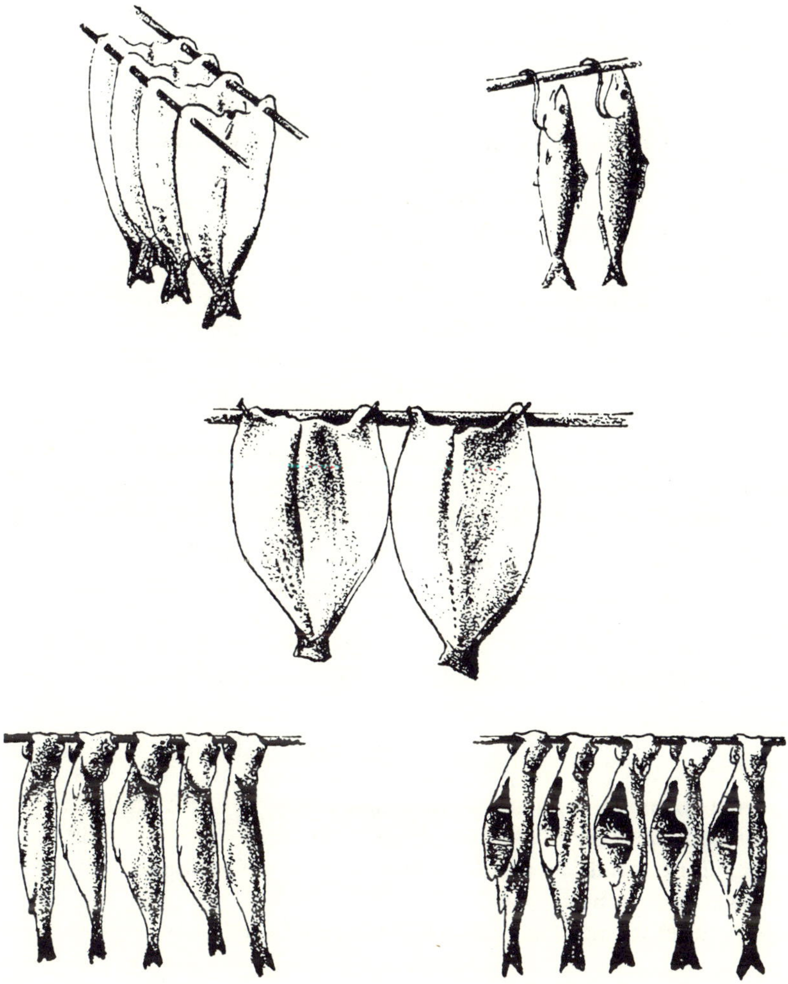

Figure 30. – Methods of hanging fish for smoking.

watched carefully, as it has a tendency to flare up and get too hot. This applies also to coconut husks or leaf ribs in tropical or subtropical climates. Ash, beech, birch, juniper, and poplar are some of the other varieties of wood used in the United States.

Some woods are preferred as fuel because they give an aromatic or otherwise especially desirable flavor to the smoked fish. Sweet bay is used in southern states for this reason, and hickory through the Middle West and the South. The wood from old apple trees makes an excellent fuel when available, giving an especially desirable color and flavor to the fish. Crab-apple wood is used in smoking oysters. The smoke from dead orange trees is reported to give a very attractive color.

In addition to suitability, the price and local varieties readily available are important factors in determining the types of fuel used for smoking fish in different parts of the world. In the British Isles, oak is believed to give the best color and flavor but as it is expensive and not always available, beechwood chips and sawdust are used. Peat and softwood sawdust are used in Scotland, especially for finnan haddie and smoked fillets. Beech and birch are used for smoking in Norway; however, in packing brisling (Norwegian sardines) oak is used exclusively for smoking the fish before canning. Green alderwood is used in Denmark, with oak or beech sawdust to smother the flames when a large volume of smoke is required. Oak wood, usually the small branches, chips, or sawdust, is favored in Holland. If oak is scarce, ash, birch, or poplar may be used. Mahogany sawdust from wood-working plants is a preferred fuel, if available.

Much also depends on the knowledge and care shown by the curer in the operation of the smokehouse. Stevenson (1899) warns that:

> "The smoke-house should always be warm and dry before the fish are put in as the development of steam is apt to injure the fish. Even when using the same kind of wood, the length of time required to smoke an article of uniform grade depends largely on the condition of the weather, much longer time being required when the weather is sultry than when it is clear and windy."

Hanging Fish for Smoking

Various methods of hanging fish in the smokehouse are used, depending both on type of fish and on local custom. The fish may be hung on one or more S-shaped iron hooks, which are in turn hung over sticks running from one side of the smokehouse to the other. If whole, the fish may be hung on round, wooden sticks inserted under the gill flap and through the mouth. When these sticks have been hung with fish they are suspended from one side of the smokehouse to the other. If the fish are split, the smokesticks may be 2-inch square sticks. Nails

If the fish are split, the smokesticks may be 2-inch square sticks. Nails are driven through the two sides of the sticks with the points projecting at a 45 degree angle. The intervals between nails depend on the average size of the fish smoked. The sides of fish are hung on adjacent nail points just below the bony neck plates, thus holding the fish open so that all of the flesh surface will be smoked. The fish are hung staggered on opposite sides of the sticks so that no two sides are exactly opposite each other. Another method is to run 1/2-inch iron rods through the fish just under the hard bony plate at the neck, one rod on each side. Thus, each fish hangs from two rods (figure 30). Twelve or more more fish may be hung on a set of 2 rods 4 feet long.

Fillets may also be hung over three-sided sticks of wood which in turn rest upon the sticks at each side of the smokehouse. Chunks, cutlets or other small pieces, and shellfish such as shrimp or oysters, are placed on wire mesh-bottomed trays having a wooden frame. These are also used for fillets. Trays are easier to load and unload but the fish may show markings from the wire mesh. On the Pacific coast the bacon hanger is widely used. It was first used in cold-smoking salmon but now for other fish. The bacon hanger is described under *Salmon Smoking.*

Chapter Eighteen

Herring Smoking

The Atlantic herring (*Clupea harengus*) was probably the first fish to be smoked on a commercial scale. Hard-smoked herring was one of the few food products in the Middle Ages which could be preserved for more than a short time with any degree of success. In an age when commerce was generally local, it found a market in many parts of Europe. As far back as the time of Edward I, Yarmouth was already noted for its smoked herring, and English merchants were exporting smoked herring to the Continent.

Herring is today the most important fishery product in the world and is about the only one not regarded as a luxury article. The most important herring smoking centers are in England, Scotland, Holland, France, and Norway, in Europe; the provinces of Nova Scotia and New Brunswick in Canada, and the States of Maine and Massachusetts in the United States. More herring is smoked in Scotland than in any other country in the world.

The four important methods of smoking herring result in widely different products. Herring that have been heavily salted and given a long cold smoke until they are hard and dry are called "hard-smoked" or "red" herring. Red or hard-smoked herring has lost much of its popularity but is still cured in considerable quantity and is the source of boneless smoked herring. These are smoked fish which have been skinned and boned. Although hard-smoked herring has better keeping qualities, a lighter smoked product is now favored. Round herring which have been lightly salted and dried in a short cold smoke are known as "bloaters." Bloaters have a short period of preservation. In the United States, only the large lightly smoked herring are classed as bloaters, but in Europe, fish of almost any size may be given this cure. Kippered herring are fish that have been split and gutted, lightly flavored with salt and given a short cold smoke. "Buckling" or "pickling" are large, fat, round herring, lightly salted, given a short cold smoke, followed by a hot smoke which also cooks the fish.

Scottish immigrants of the late eighteenth century are said to have been responsible for the introduction of "red" herring to this continent. The curing of bloaters in America is reported to have begun in 1859 at Boston, Mass., where it was introduced from Scotland. The preparation of kippered herring is believed to have first occurred on this continent in Canada about 1890. According to some reports, "buckling" were first cured in America by early settlers from the Netherlands. Other sources state that this method was brought in by German immigrants about the middle of the nineteenth century.

Hard-Smoked Herring

The herring were formerly scaled before they were unloaded from the boat. This was done by having the fishermen shuffle through the pile of fish without lifting their feet, thus stirring the heap around. This step has been almost entirely discontinued since the close of the nineteenth century, as it was found to result in a large number of bruised and belly-broken fish. Also, the amount of handling necessary in smoking removes nearly all the scales without any special effort.

Either fresh or salt herring may be used in smoking hard or "red" herring. Fresh herring are preferred since a higher quality product is obtained. In Maine, the herring are obtained principally from weirs in the Passamaquoddy Bay region. Many of the weirs are owned by groups of farmers with water-front property. They cooperate in the construction and operation of the weir. The catch is divided among the owners, who may sell to the operators of large smokehouses near Eastport or Lubec, or may cure their shares individually, as they prefer. Some fresh herring are also imported from various points in New Brunswick.

The first step in curing hard-smoked herring is brining or pickling, if the fish are fresh. This may be done by mixing the herring with dry salt, the method favored in Europe. In America, the fish are given a light salting in brine. Fresh herring intended for smoking are then taken to the salting shed of the smokehouse as soon as they are landed and placed in tanks with a capacity of about 4 hogsheads of herring. A hogshead will hold 1,200 pounds of fish. After the tank has been partly filled with 40° salinometer brine, enough herring are put into the tank to form a layer on the bottom. Approximately 30 pounds of salt is thrown on this layer of fish. Another layer of fish about 2 barrels in amount, then goes into the tank, to be covered with a somewhat heavier layer of salt. This amounts to from 60 to 90 pounds. A third or top layer of fish is added and covered with a heavier layer of salt, amounting to from 180 to 300 pounds. The top layers are more heavily salted in order to prevent the herring from being unequally cured. This would occur if the salt was equally distributed, as the brine at the top would then become weak. The amount of salt used depends on the condition of the fish and the season of the year. If the herring are large or fat, or if the weather is warm, more salt is required. In cool weather and if the fish have just been caught, less salt than usual is needed. The time required to brine the herring is also dependent on the factors just mentioned. Small herring are salted from 24 to 36 hours. Large herring are salted up to 48 hours or even longer if they are very fat.

When the herring have been sufficiently salted they are removed from the brining tanks by dip nets, locally called "wash nets." The net full of fish is rinsed around in the brine of the tank to clean the herring and is then emptied on the stringing tables. The brine drains off while the fish are being hung.

If salt herring are to be smoked, the first step is to soak them in sea water to remove the excess salt. The herring are soaked or freshened from 12 to 15 hours. Some smokers change the water at least once while the fish are being freshened; others do not.

The smoke sticks on which the herring are hung are thin, wooden rods about 3/4-inch in diameter, 3 1/2 feet long, and pointed at one end. To string the herring, the back of the fish is held in the right hand. The left gill cover is raised by the right thumb and the pointed end of the stick is inserted through the gill cover and out at the mouth. The fish are shoved along the stick until it holds from 25 to 30 herring. A workman can string from 500 to 1,000 sticks in a day (figure 31).

When a stick is filled with herring it is rinsed in a tub of clean sea water and hung on a "herring horse." This is a rectangular frame resembling a rather crude stretcher with fixed legs. It is used to transport the sticks of herring to the smokehouse and as a rack for draining and drying. A horse will hold 45 sticks or about one barrel of herring. Each frame is carried out into the open air, weather permit-

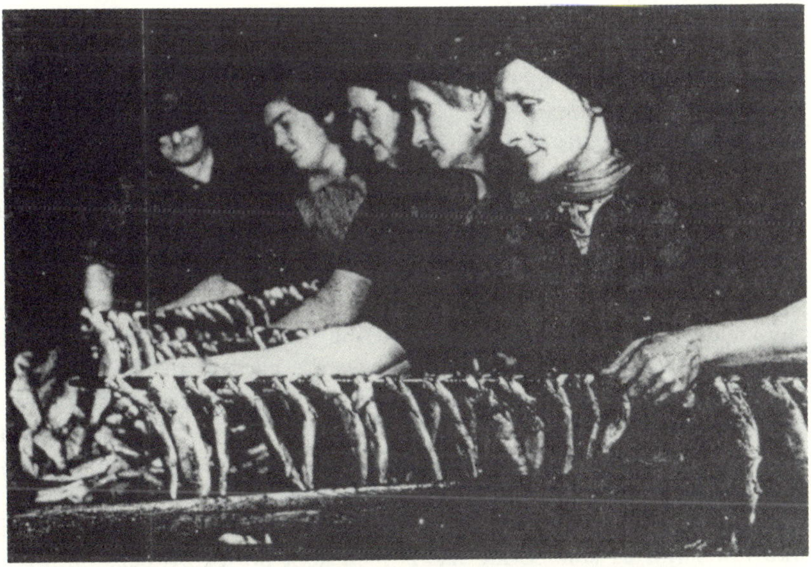

Figure 31.—Smoking bloaters showing stringing of herring on smokesticks.

ting, and allowed to stand until the water is drained off and the fish are dry enough to hang in the smokehouse. The weather determines the length of the drying period, which varies from 1 to 5 hours. The preliminary drying hardens the gill covers so that the fish are not so apt to fall from the sticks during smoking. The fish also smoke more evenly and have a better appearance. If the weather is unfavorable, that is, foggy or rainy, the herring are hung in the smokehouse after draining for a few minutes. The smokehouse doors and vents are opened and the fish are dried over a clear fire.

Each smokehouse is divided by scantlings into a number of sections called "bays." These divisions are just far enough apart so that the ends of the smoke sticks will rest on the crosspieces or stringers fastened horizontally to the scantlings. The herring are usually strung and dried in the morning and hung in the smokehouse in late afternoon. The smokehouse is not completely filled at one time. Herring suitable for smoking are often obtainable only in comparatively small quantities, and even if large amounts are available, the smokehouse is filled gradually, in order to make sure that the fish are well dried and that they will be more uniformly colored and smoked. If the smokehouse were completely filled with fish at one time, the air would become so saturated with moisture that the rate of drying and smoking would be delayed to such an extent that the fish might spoil.

At least two men are required to hang the herring in the smokehouse. One man stands in the bay with his feet on the beams, while the other stands on the ground and hands up the sticks of herring, two at a time, keeping the sharp end of the stick upward, so the herring will not slip off. The lower rows of the bays are usually filled first and the sticks are spaced far enough apart so that the herring do not touch each other. When the day's lot of herring has been hung in the smokehouse, the fires are lighted and the fish are given a preliminary smoke from 12 to 15 hours, or until the herrings are colored by the smoke. At the end of this time the fires are allowed to die down and the partly cured herring are transferred to a place nearer to the roof of the smokehouse. Another lot of fish is then hung in the lower part of the smokehouse and the preliminary smoking is repeated. The process is continued until the smokehouse is completely filled. For a large smokehouse this may require 2 weeks, while a small house may be filled in a few days. A smoldering fire is then kept up for 3 weeks until the fish are completely cured.

The fires for smoking the herring are built on the ground in the smokehouse and are usually spaced at equal intervals over the entire area, although in some of the smaller houses a low, continuous fire is built, following the lines of the back and two side walls. Almost any kind of wood except pine may be used as fuel, but white birch is generally preferred. Driftwood which has been soaked with salt water may also be used. The principal requirement is to have wood that will burn slowly and give off a good deal of smoke. The fires are kept burning continually night and day, and very slowly. The temperature should preferably be less than 70° F. If the smokehouse becomes too hot the fish may soften and spoil. The standard test to determine the degree of heat is to stretch out an arm between the rows of herring. If the air feels distinctly warm on the hand, the smokehouse is too hot. The more modern smokers have installed recording chart thermometers.

After being smoked, the herring are allowed to hang in the smokehouse until they are to be boned or otherwise prepared for shipping. If it is not possible to begin packing in a short time after the smoking process is finished, the smokehouse doors are kept open during the day, and if the weather is damp, low, clear fires are kept going in the smokehouse, to prevent the fish from absorbing moisture.

The herring are taken from the smokehouse to the packing shed where they are packed and graded as to size and quality. According to

Tressler (1923), the most important grades of hard-smoked herring are medium- scaled, lengthwise, and No.1. The best grade is medium-scaled, which is separated into two sizes, known as large and small-medium. Medium-scaled herring are packed crosswise in wooden boxes. A box of the large medium-scaled herring holds from 30 to 40 fish and the boxes of the medium-size average 45 to the box. The lengthwise herring are larger than the medium-scaled and were given the name from being packed lengthwise in boxes, which contain from 15 to 18 smoked fish. The No. 1 herring are the smallest and lowest priced grade of those regularly packed; each box contains from 60 to 75 fish. Other grades of hard-smoked herring less generally recognized are tuck-tails and magdalens. According to a leading fish-curer, smoked herring are sold in boxes of 25 to 50 fish per box. The larger fish are called No.1 and the smaller, No.2. At one time some herring were packed 100 to the box, but this is rarely done today.

If boneless smoked herring are to be prepared, the smoked herring are dumped on wooden tables; the heads, bellies, and tails are clipped off with scissors. The trimmed fish are then taken to other tables where they are weighed. Women and girls then remove the skin and bones, using their fingers and a knife. The more expert workers can skin and bone from 100 to 150 pounds of herring daily. The skinned and boned herring are cut into small strips which are usually packed in light wooden boxes lined with paper and holding from 5 to 10 pounds. Some of the boneless smoked herring are packed in vacuum sealed glass tumblers holding from 5 to 8 ounces and in tin cans to about the same weight. The containers are not heat-sterilized. Preservation depends entirely on salting, drying and smoking. Boneless smoked herring is also packed in small cellophane or glassine envelopes holding 1 or 2 ounces.

"Red" Herring

Great Britain and the Netherlands were formerly large curers of red herring. Owing to changes in food habits and to better methods of food handling, the demand for the hard-smoked or red herring has decreased but it is still prepared in considerable quantity, especially for export. The method of curing red herring followed in England and Scotland (Duthie 1911), is briefly as follows: Fresh herring are roused (mixed) well with salt, then packed flat in barrels with plenty of salt, or about 30 pounds of salt to 100 pounds of fish. The herring are not gutted. The barrels are allowed to stand on end for 2 or 3 days after which they should be filled. headed, and laid on their sides. Brine should be poured into the barrels through the bunghole until the containers will hold no more. The herring are left to cure for at least 10 days, though some curers may leave the herring in salt for as much as six weeks. On the other hand, some curers start smoking after only 3 days in salt. Duthie states that curing may be done in the tanks used for salting bloaters. If this is done the herring should be well roused on the floor with salt and turned over during rousing with a wooden shovel. When herring and salt are thoroughly mixed, they are shoveled into the tank, with more salt thrown among the herring as they are filled in, and a strong brine added after filling until the herring float. Curing in barrels is believed to be the more satisfactory method.

When the fish are considered to be sufficiently cured, they are strung on smokesticks, as described in curing hard smoked herring. Each stick is about 56 inches long and holds an average of 25 fish. The sticks of herring are spread across long narrow tanks of water, about 4 feet wide so that the fish hang in the water. The herring are allowed to steep in these tanks for some 36 hours, the water being changed once or twice during the steeping period. The object is to soak out the surplus salt.

After the herring have been freshened sufficiently they are taken out to drain and dry. This is done on racks in the open air, if the weather is favorable. If damp or rainy, the herring are hung in the smokehouse at once, but are allowed to drain or drip from 12 to 25 hours before the fires are started. The same smokehouses used for kippered herring are generally used for smoking red herring today. When the herring have drained and dried sufficiently, a low fire of billet-wood (small chunks) is started and allowed to burn overnight or for about 12 hours. The fire is then allowed to die down and the fish cool for 24 hours. A low fire of billet-wood is again lighted and kept up for 24 hours. The fire is then allowed to die out for 24 hours. When it is relighted for the third time the fire is largely of chips and sawdust in order to obtain as much smoke as possible. The fish are smoked on alternate days over a period of from 3 to 6 weeks (depending on whether the herring are intended for domestic or export trade) before they are considered cured. The temperature of the smokehouse fluctuates between 57.2° F. (14° C.) and 66.2° F. (19° C.), averaging about 65° F. (18.3° C.).

Red herring are packed in small boxes and kegs or half-barrels. The red herring barrel is a wooden-hooped, dry-ware cask like those used for grapes imported from Spain in prewar years. The heavily smoked fish (5 to 6 weeks of smoking) are known as "ham-cured reds" and "black herring." Fallen fish or headless herring are spiked on the nailed sticks used for kippered herring and smoked with the others. They go into the trade under the name of "red tenters" and "plucks."

Red herring are graded as follows:

> Best selected reds, in kegs (80 to 100 fish).
> Best selected reds, in boxes (40 fish).
> Second quality reds in boxes.
> Second quality reds in half barrels (189-200 fish).
> Red tenters and plucks in half barrels.

Bloaters—American Method

Most of the bloaters cured in the United States are prepared from large, salt herring. Round or ungibbed herring are preferred, although gibbed fish are used when herring are scarce. The salted fish are soaked in large, square tanks of fresh water from 12 to 24 hours, the time depending on the saltiness of the herring. If the bloaters are made from fresh herring, the fish are held in a 90° to 95° salinometer brine for an average of 48 hours, then rinsed and hung on sticks. The salt herring, after soaking, are dipped onto a stringing table and strung on smokesticks as in the curing of hard or red herring but with fewer fish, generally 15, hung on a stick (figure 31). In some places the herring are soaked for an hour or two, hung on smokesticks, and then given a second soaking for 15 or 16 hours.

The herring are drained for a few minutes after soaking, then hung in the smokehouse, and the smoking process is begun, with ventilators and drafts of the smokehouse left open. Stevenson (1899) states that:

> "In order to 'bloat' the herring must be thoroughly moist, and after they have commenced to dry in the smokehouse the heat must be increased. If they are permitted to hang 10 to 12 hours without heating they will not bloat, but will become hard herring. The smoking is continued from 2 1/2 to 6 days, when the fish are usually sufficiently cured."

On the Pacific coast the herring are smoked and dried over a low fire of alderwood at a temperature of 65° to 70° F. for 48 hours. At the end of that time a smudge is built by burning low heaps of sawdust to obtain as dense a smoke as possible with little heat. The temperature of the smokehouse should be about 60° F. This smudge is kept up for from 2 to 5 days, depending on the weather. In damp or unfavorable weather about 7 days are required to cure the fish from the time they are placed in the smokehouse. Bloaters are packed in light, wooden boxes holding 25, 50 or 100 fish with a net weight ranging from 25 to 35 pounds, and in smaller boxes with a net weight of 10 pounds.

Bloaters – English Method

The method of cure followed in the British Isles differs to some extent from that used in the United States. According to Duthie, "the bloater trade is more an English than a Scotch industry." The herring are brought in fresh from the drifters (gill netters). Trawl caught herring are not used if possible, as herring taken by otter trawl tend to be more or less bruised and damaged. The first step is to wash off the herring, then dry-salt them in Spanish salt, using enough salt to cover the fish and mixing both salt and herring well together. About 25 pounds of salt are used to 100 pounds of herring. Salting is considered a very important step, as herring too heavily salted will not have the flavor demanded by the trade. Duthie (1911) states that the herring are left in the salt "overnight," that is, from 12 to 14 hours. English curers have informed the author that if the herring are to be packed as bloaters they should not be salted more than 6 hours for a first-class product. Some smokers have large tanks in their plants to store herring in brine. Herring are salted in tanks to be held until the herring season is over, when they are withdrawn as needed and freshened in water 12 to 16 hours before being hung on smokesticks.

The herring are rinsed free of salt and hung on wooded smokesticks as for hard or red herring. From 12 to 16 herring are hung on a stick. The fish should be well spaced, not touching each other. The sticks are hung in the smokehouse and the herring allowed to drain and dry overnight, or up to 24 hours, the time depending on the rush of business. A fire is then built in a thin, even layer over the entire floor of the fire pit. It is made of hardwood chips and sawdust, beech, birch, or other wood; one or two firms prefer oak. Duthie states that bloaters are smoked in much the same way as kippers and that a fire of hardwood billets is usually preferred to chips and sawdust. The soft fuel gives more color than is desirable, as bloaters should be flavored rather than colored in the smoke. From 8 to 12 hours of smoking at a temperature of about 85° F. will generally cure the herring sufficiently for the British market. The product is allowed to cool and it is then packed in boxes, usually of one stone (14 lb.) each.

Kippered Herring

Kippered herring is the most important cured fishery product in Great Britain. It is not so popular in the United States. This may be ascribed, in part at least, to the fact that in the past herring of small size and poor quality have been kippered in an effort to dispose of otherwise unmarketable fish. The demand for kippered herring is slowly increasing in the United States; and a well-prepared product, properly merchandised, should eventually find a considerable market.

The best kippered herring is supposed to come from Scotland and the Isle of Man, but considerable quantities of kippers are cured in many of the fishing ports throughout Great Britain. A well-prepared kippered herring makes a tasty breakfast dish. Almost every firm differs as to one or two details of the curing process for kippered herring, but in general the process is as follows: Fresh herring only are used for kippering. The fish should be of fair size and in good condition. A reasonably good fat content is desirable.

After being washed to remove slime, blood, scales and other debris, the herring are split down the left side of the backbone from the tip of the head almost to the tail and are then opened to lie flat in one piece. Gills, viscera, blood and membranes are cleared away, and the fish are washed well in fresh water. After draining for a few minutes, the herring are immersed in a 90° to 95° salinometer brine. The time required for brining averages 20 to 30 minutes. The principal factors deciding the length of the brining period are the amount of fish in a batch, the condition of the fish, as soft herring require more brining than firm fish, and the market where the product will be sold. Some markets like a saltier product than others. The length of preservation desired must also be considered.

The herring may be dyed at this point, or the dye may have been previously added to the solution in which the fish were brined. If added to the brine, the amount of dye is regulated by the length of the brining period. If not, the herring are dipped for about 30 seconds in a dye solution with a strength of about 1 to 2,000, that is, 1 gram of dye to 2 liters of water. The dye used is either annatto or a similar compound. Curers state that a good dyeing can only be learned by practice and like brining, varies, according to the weather, condition of fish, and the amount handled in a batch. Fish curers are almost unanimous in the belief that the better quality kippers should not be dyed, but occasionally some are dyed to meet the demands of a special trade. This practice is much more general with kippered herring of lower grade.

Figure 33.—Kippered herring in smokehouse.

After draining a few minutes, the herring are hung on smoke sticks. The sticks are wooden bars 4 feet long by 1 1/2 inch by 1 inch. There are 16 nails driven in on each side, the points projecting upward at a slight angle. The nails are in pairs 3 1/4 or 3 1/2 inches apart, according to the probable size of the herring, and about 1 1/2 inches between each pair of nails, so that a stick will hold 8 herring on a side. The fish are hung by piercing the head through a pair of these nail points so the herring are open flat with no two fish touching. The sticks of herring are hung in the smokehouse to drain and dry until evening. The lowest row is hung about 12 feet above the firepit floor. Some curers use almost any hardwood chips or sawdust while others use only oak, claiming that other woods do not give the smoke flavor essential to good quality kippered herring (Duthie 1911). The fish are smoked lightly for 6 to 8 hours. One or two refuelings of the fires are usually needed, the second and third fires being lower than the first with less heat and more smoke. Some heat is necessary at first if the herring are to take color properly but the fish must not be overheated or they will be scalded and fall from the hangers (fig.33).

The kippered herring are allowed to cool and dry for several hours. If this is not done the herring are apt to sweat and arrive in bad condition. After cooling, the herring are sorted and packed in boxes of several sizes, the most popular holding a stone (14 lb.) of fish. A growing export trade is being done with the United States and the British Commonwealth. Kippered herring intended for the export trade or the best class of local trade, are wrapped in pairs of cellophane envelopes before packing in boxes.

A development in the curing of kippered herring that has resulted in an improved product was introduced by a firm of fish curers in Fleetwood, England, some years before the outbreak of World War II. This is the boneless kipper. The herring are dressed mechanically (fig.34). They are headed, the thin part of the belly is cut off, and the blood, viscera, and backbone are removed by a machine handling around 2,000 fish per hour. There is no difference in the brining and smoking procedure. The boned kipper must be handled more carefully than fish dressed in the ordinary way but the product is of high quality and is practically free from bones and waste. In a number of other packing centers, boneless kippered herring are being prepared by hand. The author has put up several experimental hand-dressed packs and finds lit-

Figure 34.—Machine splitting, eviscerating, gilling, and cleaning herring in preparation for kippering.

tle difficulty in using this method. Loss in dressing is slightly greater but the improvement in product more than balances any loss in weight. The boneless herring may be dressed almost as quickly as by the present method for the non-boned product.

The method for kippering herring in the United States differs but slightly from the British cure. At least one of the important curers follows the British cure exactly. The principal differences are that the fish are given a heavier brining and a longer smoke cure, that is, from 12 to 16 hours. Since this country is larger and the demand here is rather uncertain, a more durable product is required. Even then kippered herring will not keep long at ordinary temperatures. The product is packed in 10-pound wooden boxes and is usually stored at temperatures of 33° to 35° F. Experimental freezing of kippered herring has not been particularly successful. Studies at the University of Washington College of Fisheries indicated that the best results were obtained by freezing fresh fish, to be thawed and kippered as required. The smokehouse developed at Halifax laboratory of the Fisheries Research Board of Canada has been found suitable for smoking kippered herring commercially in Canada. It may also be used with mackerel and most other smoked fish, according to information from Canada.

Buckling

A small amount of herring is smoked as buckling in the United States, principally in New York. Buckling is an appetizing smoked fish product and deserves a much wider market. It is believed to be of Netherlands origin but is one of the principal smoked fishery products of Germany. Fresh, fat herring of medium size are preferred in the preparation of buckling. The fish are rinsed in fresh water to remove blood, slime, scales, and any other debris. After draining, the herring are left in a 70° to 80° salinometer brine from 2 to 3 hours, the length of time depending on the strength of the brine and size of the herring. In some districts the fish are not cured in brine but mixed with dry salt and left in a heap for about 2 hours or until sufficient salt has penetrated the herring.

The herring are rinsed in fresh water, then strung on iron or wooden smokesticks or spits. The method followed in stringing is that used in stringing hard-smoked herring or bloaters. Each stick is about a meter (3.3 ft.) in length and holds from 15 to 25 herring, the number varying according to the size. Care must be taken to space the fish so that no two touch. The sticks or splits are laid on wooden frames holding from 15 to 25 spits. These frames are slid into movable or stationary racks, where the fish are left to drain. In some cases when the weather is favorable, the herring are also given a sun-drying in addition. This increases the capacity of the smokehouse as a greater quantity of fish can then be cured during the working day.

When the fish are properly drained, the frames are placed in the smokehouse. A fire is built to burn with a bright clear flame, without much smoke. Sticks of alder, beech, or oak are the fuel used. After 1 to 1 1/2 hours, the herring are considered dry. The fire is covered with alder chips to produce a heavy smoke and give the fish the golden yellow color required. If the chips do not give off sufficient smoke, they may be dampened occasionally by a sprinkle of water; this is also done if the fire burns up too strongly. The regulating damper at the top of the chamber is first closed half-way, and later all the way. It takes approximately 1 to 1 1/2 hours to complete the cure. The whole process requires about 3 hours. The quality of the product is said to depend largely on the size and quality of the herring and the care with which the fires are regulated. The buckling are marketed in boxes of 5 to 14 pounds. Herring split down the back like kippered herring and then given the buckling cure are known as "fleckheringe."

Modified Buckling Cure

The writer has found that the following process, a modification of the buckling cure, will produce an appetizing article suitable for distribution in the United States. Fresh or frozen herring may be used. Frozen herring should be thawed out overnight. The herring are covered with three-quarter ground salt, using about one part of salt to four of fish. They are allowed to cure from 2 to 3 hours, and are then rinsed off in fresh water. After draining, the fish are given a drying from 2 to 2 1/2 hours in an artificial drier at a temperature of 60° F. preferably, and not more than 70° F. All surface moisture should be removed by this time. After being strung on rods, the herring are given a cool smoke at a temperature no higher than 70° F., for 6 hours. The fire is then built up with the result that the fish are partly cooked for 45 to 60 minutes, at a temperature of 180° F.

Chill-storage temperatures between 33° and 36° F. must be used if the product is to be kept in good condition for periods longer than a few days. It must be handled even more carefully than kippered herring.

Sprats

Smoked sprats (*Clupea sprattus*) are a favorite product in all the countries of the Baltic area, especially in Germany. Some smoked sprats or "Keiler Sprotten" were imported into the United States from Germany before World War II. A smoked sprat substitute is now cured in the United States, prepared from the Atlantic herring (*Clupea harengus*). The German process is briefly as follows:

Fresh sprats are graded for size so that each lot is uniform. They are then brined for one hour in a salt brine testing 80° salinometer. After being rinsed thoroughly, the sprats are piled in baskets to drain off the surplus moisture. They are then hung on iron spits, much as in sardine canning; that is, a thin iron rod is run through the eye sockets of the fish. When a rod is filled it is placed in a wooden or iron frame. However, the fish must not be hung too closely on the rods. The frames are stacked to a depth of 4 or 5 frames in the standard type hot-smoking oven used for buckling. The smoking process is almost the same as for buckling; that is, the fish are first cured over a bright, clear fire for about 1 hour. The fire is then covered with alder chips and the damper closed, building up heat and dense smoke. The cure is completed in about an hour, so that the entire smoking process takes 2 hours or a little more.

The smoking of sprats in the United States is confined to New York City, which is the only commercial market for this product. Some sprats have been smoked in Nova Scotia and sent to New York. The American sprats are cured from small spring herring, 5 to 7 inches in length. Some frozen herring shipped in from Newfoundland are used. The herring are first washed thoroughly, allowed to drain for a few minutes, and then brined for 45 minutes in a 90° salinometer brine. After being rinsed, the fish are drained thoroughly, then spread thinly on wooden-frame trays with bottoms of wire mesh. The herring should not be allowed to touch each other. The trays are placed in a finnan haddie smokehouse at a distance of 18 inches to 2 feet above the fire, where the fish are smoked from 2 to 4 hours. At the end of this time they should have a golden brown color and be sufficiently cooked so that they are ready to eat. These sprats remain in good condition for a few days only.

Chapter Nineteen

Salmon Smoking

Smoked salmon is prepared by several different methods from raw material of various grades and types. Formerly a large proportion of pickled or hard-salted red salmon from Alaska was shipped to the eastern part of the United States and Europe for smoking. Today, mild-cured chinook or king salmon are most used in smoking, but varying quantities of silver or coho salmon are also mild-cured. These sell at a somewhat lower price than mild-cured king salmon, and make excellent smoked salmon. Some fresh Atlantic salmon is smoked after being salted very lightly for a few hours. This product is generally known as "Nova Scotia" salmon in the eastern markets and is regarded as the finest grade of smoked salmon, but far less is used than of mild-cured salmon for the same purpose. The only smoked-salmon product requiring an appreciable quantity of fresh raw material is kippered salmon, for which fresh or frozen salmon is invariably used.

American Method

In preparing smoked salmon from mild-cured fish, the sides of salmon are taken out of the tierce and soaked overnight in fresh water, or for a period of 12 to 16 hours. The water is changed two or three times. Ten or 12 hours of freshening should be sufficient but a more thorough soaking may be required to satisfy some markets. Certain smokers freshen salmon for 10 hours in a tank of running water, especially if a large quantity is to be smoked. When sufficiently freshened the salmon are washed to remove all traces of blood, slime, or encrusted salt, then carefully wiped with a clean soft cloth.

The next step is draining and trimming. Draining is often done by water-horsing; that is, the salmon are placed in a pile, flesh side down, and a weight is placed on top to press out the water. After enough moisture has been drained from the flesh the sides are trimmed of any ragged edges and are taken on barrows or hand trucks to the smokehouse.

Wire hangars are used for hanging the salmon on sticks in the smokehouse. These hangers are made of steel wire or light iron and have six points at the lower end at right angles to the frame and a curving hook at the top, to hang over the smoke stick. A side of salmon is laid out flat, skin side up. The points of a hanger are pressed through the skin at the nape of the neck end. The side is then passed to another workman in the smokehouse, who hooks the handle over a round smoke stick. In hanging the salmon, care is taken to leave sufficient space between sides to guard against the crowding of the fish together or overloading the smokehouse. If this is not done, the salmon is unevenly smoked and of inferior quality.

The time required for the smoke cure depends primarily on how long the salmon are to be kept, and on market preference. On the Atlantic coast, light smoking for 8 to 10 hours is considered sufficient. In fact, some curers dry the fish to the desired texture in an oven, using very little smoke. In other sections of the country longer cure is required because the smoked salmon may not be consumed within the next few days and the trade desires a product with a definite smoke flavor and somewhat firmer texture.

It is the opinion of some fishery authorities that the Pacific and midwestern smoked salmon are superior to the average Atlantic product. The Pacific coast method requires a longer cure, more work, and entails more expense than the Atlantic method. The following method is descriptive of western salmon smoking practices (Jarvis 1936b): After the smokehouse has been filled, a clear fire is started in the pit below and for some hours the fish are smoked over a clear fire with the ventilators left open so that moisture can escape. This prevents the salmon from sweating in this initial period of smoking, which is really more or less a drying process. A few smokers dry off the salmon using mechanical fans until a thin shiny skin has formed on the surface, before hanging in the smokehouse.

When the first period of the smoke cure has been finished, that is, after 24 to 48 hours of light smoke, with smokehouse ventilators open, the ventilators in the top of the smokehouse are closed and the fire is smothered with sawdust. A dense smoke is thus created in which the salmon are cured for an additional period that requires 48 to 72 hours. For a still more durable product, one that may be marketed, over a wide area and will keep longest under average conditions of temperature, handling and storage, the smoke cure may require a week. In this case, the fire is kept low and smouldering during the entire period of the cure. A dense smoke is not formed but the product acquires a distinct smoke flavor because of the long cure (fig.35). It is as much a process of dehydration as smoking.

When the cure is completed the smokehouse doors and ventilators are left open. After the smoked sides are sufficiently cooled, they are weighed, wrapped in oiled or parchment paper, and packed in light wooden or fiber boxes with an average net weight of 30 pounds. Smoked salmon must be kept in chill storage at temperatures of 34° to 40° F. if they are to be held any length of time, especially in summer. A well smoked side of salmon should have a smooth, glossy surface with a light brown color. The flesh inside should have the texture of smoked ham.

The length of the smoking period and other factors involved in smoking salmon vary according to locality, type of product demanded by the trade, temperature used in the smoking process, atmospheric temperature, humidity, and similar factors. The process must be altered to meet changes in these conditions. Exact data as to temperatures, length of time, density of smoke, and other factors required to give the best results are lacking. This is a cold-smoking process, however, and though the fire must be sufficient to dry the salmon it must not give off too much heat or the fish will be partly cooked, so that it will not have the desired texture and flavor and will soon spoil. The temperature should not exceed 90° F. and in general should be lower, or about 75° F. if atmospheric humidity is at all high. As to the best type of fuel, alder wood is favored and is most commonly used on the Pacific coast but almost any non-resinous wood, such as maple or beech, gives satisfactory results. Oak and hickory are favorite fuels among salmon smokers on the Atlantic coast but they are often unobtainable and many substitutes are used as fuel for smoking.

A small amount of smoked salmon is sliced like bacon or ham, wrapped in cellophane and sold in half- or quarter-pound packages to

Figure 35.—Smoking salmon, Pacific coast cure.

the delicatessen and grocery trade. Sliced smoked salmon is also packed in 4- or 8-ounce screw-top glass jars. Small packages holding 1 or 2 ounces for individual servings are mounted on cards for sale to the tavern trade. Hard smoked or heavily cured sliced smoked salmon is also packed in quarter-oil cans of the type used for Maine sardines. A little olive or cottonseed oil is added to each can which is then sealed hermetically but not sterilized. While this product is not as perishable as the usual smoked salmon, it does not have an unlimited period of preservation and should not be exposed to high temperature or other favorable storage conditions. The maximum preservation is achieved by keeping this product in a refrigerator or refrigerated showcase. Some smoked salmon in oil was imported from Japan and various north European countries before World War II. This product was also unsterilized but in certain cases such as urotropin were added. The use of this preservative and a number of others are made illegal under the U. S. Food, Drug and Cosmetic Act. Sodium benzoate may be used in amounts less than one-tenth of one percent if presence and amount are stated on the label.

Scotch Method

Scotch smoked salmon is generally known as kippered salmon but the method is different from that used for kippering salmon in America. The product also is different and most closely resembles the light-smoked salmon of the United States. There is some difference in the method of cure, two variations being given here.

The salmon are washed thoroughly, the heads removed and the fish are split down the back. The belly cavity is cleaned of viscera and other offal, coagulated blood is scraped away, and the fish again washed thoroughly to clean them of blood stains or slime. After draining until surplus moisture is removed, the fish are rubbed with a mixture of salt and brown sugar. They are packed in a tub or barrel, with more of the mixture between the layers of fish. The flesh should absorb enough of the salt-sugar mixture for smoking purposes in 48 hours. The fish are then rinsed and three small double-pointed sticks, with a length approximately equal to the width of the fish, are fixed through the skin, one near the napes, one about midway and the third near the tail end. The sticks are intended to keep the fish spread out during smoking. The salmon are then air-dried for a few hours, after which they are hung in a kippered herring or finnan haddie smokehouse and smoked for 10 to 14 hours. This is a very mild cure and the salmon remain in good condition for only a few days at ordinary temperatures (Duthie 1911).

In the second method, the fish are first washed thoroughly, then the skin is scored slightly with a knife at intervals of about 2 inches to penetrate the skin but not to cut into the flesh. The scoring is done longitudinally. The salmon are then split in sides and cleaned of all viscera and other offal. Any blood remaining in the veins around the belly cavity is gently pressed out from the flesh . The sides are then washed thoroughly in ice-cold water, where they are allowed to remain for about half an hour to make the flesh somewhat firmer. The sides are drained to remove surplus moisture and trimmed of any ragged bits of flesh or pieces of membrane.

A mixture is made in the proportion of 6 pounds salt, 4 pounds brown sugar, 1 ounce saltpeter, and 1/2 ounce each of bay leaves, black peppers, cloves, coriander seed, and allspice. The spices are ground or crushed and mixed well with the other ingredients. The sides of salmon are dredged in this mixture which is also rubbed into the flesh. The sides are picked up with as much of the curing mixture as will cling to the flesh and packed in a tub, scattering a small amount of the mixture between layers. After 48 hours the sides are taken out of the tub, scrubbed and soaked in cold fresh water for 1 hour. The sides are then laid in a small pile, flesh side down, with weights on top to press out moisture. They are left here several hours.

The sides are then hung on smoke sticks and dried in the open air until a thin skin or pellicle forms on the surface. A fan or blower could probably be used for this purpose. When the salmon are sufficiently dried they are hung in the smokehouse where they are smoked over a low fire of beech or other white wood-chips for 16 to 24 hours. A dense smoke is then built up and the fish are smoked in this for an additional 8 hours. At no time should the temperature in the smokehouse be more than 80°) F. This smoked salmon has a greater length of preservation than the first but cannot be expected to remain in good condition more than 10 days or 2 weeks at room temperature.

Norwegian Method

The following description of the Norwegian method of smoking salmon was given by Duthie (1911): "Cut the head off, and split the fish down the back. Was it clean, and then put it in salt or ordinary pickle. After lying in the pickle for 3 days, the fish is taken out and washed in clean, fresh water and then stretched upon pieces of lath. These pieces of lath are about 1 1/2 inches broad, but quite thin. They are cut to a length corresponding to the breadth of the fish, and sharpened at the end. One of these spits is put across the back of the fish at the lugs or shoulders, another half-way down, and if the fish is large, another further down, the points of the spits being stuck through the skin of the fish. The fish is then "tentered" and hung in a chimney, where it is smoked over a fire of fir branches for 1 1/2 to 2 days. Salmon cured by this process are somewhat similar in appearance to an Aberdeen-cured haddock, but rather darker in the color. They are cut into collops and sold for immediate use, generally at high prices.

> "Salmon meant to be kept for 2 or 3 months are much harder cured. The fish are split into halves to facilitate the operation of curing, and to make them easier to handle and pack. The barrels for packing should be clean and tight, and preferably of hard wood. It is advisable first to rub the fish well with a mixture of brown sugar and fishery salt, as in the Scotch method, and a little of the same mixture might be thrown in between the pieces of fish as they are packed into the cask. Hard packing should be avoided; sufficient room should be left to let the pickle circulate freely. The barrel

should be filled full of strong, clean pickle (which in this case had better be filtered) the end put in, and the barrel laid on its bilge. To properly cure, salmon treated in this way should have the pickle poured off and fresh pickle substituted, the fish being well washed in clean pickle before being repacked. As a precautionary measure, the fish might be taken out, washed, and repacked after 2 or 3 weeks in cure, even if there were no suspicion of staleness. When required, the pieces of fish should be taken out, well washed in clean, fresh water, and smoked in the same way as the milder-cured fish. Owing to the rich nature of the fish, this is the only method by which it seems possible to preserve salmon for any length of time apart from canning. Dry-curing would fail to preserve such a fat fish. In Norway, all the smoking is done in the chimneys of dwelling-houses in much the same way as finnan haddie curing was originally done on the Kincardineshire coast."

Experiments in the Norwegian method as given above indicate that a few revisions would improve the product, at least for this market. Smoking with a fire of fir branches was found to give a decidedly bitter taste. The flavor of the fish was much better when a low fire of green alder or some hardwood was used. The fish smoked much better and with a more even color when they were dried in a current of air until a pellicle or film is formed on the surface, before putting them in the smokehouse.

German Method

Fresh Rhine River salmon was the chief source of raw material for salmon smoking in the nineteenth century. A small amount of Rhine salmon is still smoked but most of the salmon used for smoking during the past 40 years has been mild-cured salmon imported from the United States. Small additional amounts of fresh or frozen salmon have been imported from Norway.

The first step in smoking mild-cured salmon is to freshen them in running water. Mild cured salmon are freshened from 5 to 10 hours. In winter, freshening takes longer, about 15 hours for mild-cured or 30 hours for heavily salted salmon. Frozen salmon are thawed in cold water, cleaned thoroughly but not scaled, then are split into sides or split open to lie flat in one piece and placed in enough salt to cover.

They are left in salt from 2 to 3 days, then rinsed and set out to dry in the air. Mild cured salmon are also set out to dry when they are sufficiently freshened.

The salmon are inspected either just before or just after drying, and any sides showing cracks or breaks in the flesh are given a protective coating to prevent additional breaks and to keep the flesh together in a single solid piece. This also permits the sides to have a more attractive appearance after smoking and they may readily be cut in thin slices. A solution of isinglass is often used for this purpose. Another coating is composed of a mixture of gelatine, glycerine, powdered pumice, and chemically purified fish glue (Hoffman; undated).

After drying for several hours, or when no surface moisture is apparent, the sides are taken to the smokehouse. Several different types of salmon smokehouses are used but nearly all are of the oven type used in the curing of bacon. A typical smokehouse is 4 meters (13.1 ft.) high. The salmon are hung at a minimum distance of 2 meters (6.6 ft.) from the fire. During the first part of the smoking process the fish are cured over a smouldering fire of small logs, which is later covered with sawdust and chips to build up a denser smoke. It is believed that oak, alderwood, or mahogany sawdust are the most suitable, giving the best results.

Another type of salmon smokehouse is a chamber 1.5 meters (5 ft.) wide and 3.5 meters (11.5 ft.) high and long. In one corner there is a door with dampers to draw off the smoke and steam. Two laths, one on each side, run the whole length of the room at a distance of from 6 to 8 inches from the ceiling. The ends of the smokesticks on which the salmon are hung, rest on these laths. A heap of oak chips is built on the middle line of the floor, for the length of the room. This is covered with oak sawdust and the whole smothered with ashes. The fire is ignited in the evening at the end near the door and during the night gradually burns the whole length of the room.

The temperature in either of these smokehouses should be about 20° C. (68° F.) or at the highest 25° C. (77° F.). A higher temperature makes the salmon flesh soft and pulpy, the flavor is not so good, and the flesh is difficult to cut. The fish are smoked from 3 to 5 days in either type of smokehouse, depending on the size of the fish and length of time for which preservation is desired. When the smoking is finished the sides of salmon are packed in shallow individual boxes lined with parchment or oiled paper. Three or four such boxes are packed in a tin-plate container which is afterward soldered shut.

Hard-Smoke or Indian-Cure Salmon

Some attempts have been made on the Pacific coast to prepare a heavily smoked and dried salmon known variously as hard-smoke, Indian cure or, in Alaska, "beleke". Although it is superior in keeping quality and equal in flavor to salmon smoked by other methods, it has not met with much favor outside Alaska and the Puget Sound region as it is dull in color and hard in texture. Hard-smoke salmon makes an excellent appetizer or relish to be served with beverages, and there are possibilities of developing a better market. This product most closely resembles jerky (charqui) introduced to our western region by the early Spanish colonists. Red and coho salmon are the species generally used in preparing Indian-cure salmon. Other species of salmon may be used, but only occasionally, when the more desirable types are lacking. Cobb (1930) states that the backs only are used which are cut into two or three long strips. The bellies are pickled and sold as salted bellies. Curers of beleke (hard-smoked salmon) have informed the author that, although this may be done, it is quite usual to smoke the whole sides of salmon by this method.

If the bellies are to be utilized as pickled or hard-salted salmon, the remaining edible portion is split in two sides, the backbone is removed and each side is cut longitudinally into several strips. The strips may or may not be washed in a light brine. In some places sea water is used; in others, fresh water. The largest, thickest strips of back flesh are then placed in a tank of 90° salinometer brine; an hour later strips of medium size are put in, and after an interval of another hour, the smaller pieces. This procedure is followed so that all sizes will have the same degree of pickle. The strips are removed and drained after a period of 16 to 20 hours. If whole sides are to be used after cleaning and splitting, as described under the preparation of pickled salmon, the fish are washed thoroughly, drained, then brined overnight or for a period of 10 to 12 hours in a 90° brine.

After brining, whole sides are fixed on hangers, while strips are usually suspended by cords, run through one end as in smoking bacon at home. The fish are given an air drying of 24 hours to remove the surplus moisture. At the end of this time the salmon are placed in the smokehouse with the ventilators left open and the salmon are smoke cured over a fire of green alderwood. The smoking is done slowly at a low temperature of not more than 70° to 80° F. Two weeks is the average time required to smoke hard-cure salmon (beleke).

This product was first prepared around Kodiak, Alaska. Hard-smoked salmon is said to have better keeping qualities than other smoked fish, remaining in good condition for more than a year. If surface mold begins to appear in storage, the fish are taken out, scrubbed in brine, given an air drying of several hours, and then smoked for 24 to 48 hours, after which they are again stored in a cool, dry place.

Kippered Salmon

Kippered salmon probably has a larger sale than any other smoked fishery product on the Pacific coast. It is sold in a few large centers in the east and middle west, but the great part is consumed in the western part of the United States. Practically all kippered salmon is prepared from white-fleshed chinook (king) salmon. This fish is not in so much demand in the fresh-fish market as other species, being considered inferior to the other salmon by reason of its paler color. It is equal to the brighter colored salmon, however, and often has a better flavor. A constant supply of fresh fish at prices that will make a profit cannot be certain throughout the year, or for any season of the year, but frozen salmon is available the year-round, giving the curer an assured supply of raw material without wide fluctuations in price. Therefore, frozen salmon are used during a great part of the year and are split before they are completely thawed. Fresh salmon, being much softer in texture, require more care and skill in splitting, and the smoking period must be somewhat longer than if frozen fish are used.

The first step in the curing process is to thaw the salmon in tanks of cold water. In some establishments this is done with running water; in others, with standing water changed several times. The time required for thawing varies from 8 to 15 hours depending on the size of the fish, total amount in the tank, and whether or not running water is used. Smaller sized fish placed in running water will be sufficiently thawed in 8 hours or less.

As the salmon have already been cleaned and dressed before freezing, they are split into sides. After the backbone has been removed, each side is cut into a number of smaller pieces. These pieces usually weigh about 1 pound each after curing and are separated according to thickness. There are usually three grades or sizes: chunks, the thicker part of the back flesh; thins, pieces of flesh not quite so thick; and strips, thin pieces from the bellies. The names used for the grades may vary with the locality and among different curers, but the separation into three grades is followed by practically all smoking establishments on the Pacific coast. The thinner pieces will cure more rapidly, which

Figure 36.—Kippered salmon, Pacific coast: Moveable racks.

is one reason for separating them, while another is that the thicker pieces are considered better grade and bring a higher price. The third grade or size, for there is little difference in the quality, usually goes to the low-price markets. Fish smokers believe that the strips are the best kippered salmon.

The cut and sorted pieces of salmon are soaked in a 90° to 95°) salinometer brine, for from 1/2 to 2 1/2 hours depending on the size and thickness of the pieces, preference of the local market for which the salmon is destined, and on the time required for shipment. If the kippered salmon is to be shipped a considerable distance it receives a heavier brining since a longer period of preservation is necessary.

When sufficiently brined, the pieces are drained and then dipped into a tank or tub of coloring matter. The dye may be added to the brine, combining the two operations, in which case the amount of dye used is less than when the fish is colored by dipping. The dye most often used is 150 Orange I, an aniline dye, the use of which is permitted under United States Food, Drug, and Cosmetic Act (Jarvis 1936b). Other red or orange dyes on the permitted list may be used. Where the fish are dipped in dye after brining, experiments conducted at the College of Fisheries, University of Washington, indicate that dipping for 15 to 30 seconds in a solution made up in the proportion of one part of dye to 3,000 parts of water is sufficient. This is given only as a general formula, to guide those without practical experience. The curer must determine his requirements by experiment, and according to the desires of his customers as to the shade of color preferred. Anderson and Pederson (1947) state that for general purposes about 3 to 4 ounces of dye to 20 gallons of water should be used. The dye should first be dissolved in a small quantity of warm water, then added and mixed in the total volume of water. A dip of 15 to 30 seconds is sufficient. The flesh is dyed because of a popular prejudice against a lightly colored kippered salmon. The dye used is harmless and does not affect the quality of the fish in any way, while it does give it an attractive color. For certain markets, principally in States where all artificial food coloring is prohibited by law, no dye is used by Pacific coast smokers. Atlantic coast smokers rarely color kippered salmon.

When the salmon has drained for a short time, the pieces are placed on wire-mesh-bottomed trays, made of half-inch mesh, with two-inch wooden frames for sides (figure 36). These trays should be thoroughly cleaned before use and the wire mesh rubbed with lard-oil or other edible oil to prevent pieces of fish from sticking to the wire. The pieces in any one tray are nearly as possible of the same size and thickness. They must not touch, or an even, sufficient cure will not be obtained. The individual trays are slid onto a rack holding several tiers of trays and moving on wheels. When filled, the rack is run into the smokehouse, or the trays may be placed directly in the smokehouse on fixed racks.

The pieces are allowed to drip for a few hours in the smokehouse, but a suggested procedure which would probably shorten this period and result in a better product would be to dry the trays of fish for an hour or two under a strong current of air at a temperature of about 70° F. When the excess moisture has drained away, the fire is lighted and the pieces are partially dried and smoked over a medium fire. The temperature in the section holding the fish should preferably be about 80° F. as a maximum, for from 7 to 13 hours. At the end of this time the fire is increased and the salmon are given a hot smoke by which they are partially cooked. Care must be taken that the fish do not get overheated for any length of time or they will be softened and spoiled.

Hot smoking or barbecuing takes 1 hour at a temperature of from 170° to 180° F. In some establishments the time is 25 to 35 minutes at a temperature around 250° F. When the fire is built up it must be regulated by means of drafts and ventilators so that the temperature will not be higher than desired.

When the smoking process is finished, the kippered salmon is thoroughly cooled before handling. In some cases this is done by throwing open the doors of the smokehouses. In other instances, where plants are equipped with movable smokehouse racks, the racks are run out on the floor and the fish are cooled under a current of air from a fan. The pieces are given individual wrappings of vegetable parchment and are then packed in a small box or basket. A container holding 10 pounds is the most popular size.

Kippered salmon is perishable, spoiling after exposure of a few days at ordinary temperatures, so if not to be sold immediately, it should be kept in chill storage at temperatures of 25° to 40° F. and retailed from refrigerated showcases. A certain amount of kippered salmon is intended for shipment to distant markets or is stored to fill rush orders. For these purposes it is frozen and held in storage for use as required. The freezing temperature and length of time required for freezing are the same as for fresh fish. As in freezing fresh fish, there is some variation in the temperatures in different plants but in a typical instance, kippered salmon is placed in the sharp freezer at -10° F. and left there for 10 to 12 hours. The storage temperature should be about 0° F.

Chapter Twenty

Smoked Haddock and other Groundfish

The smoking of haddock has declined greatly in importance in the United States during the past 20 years. At the beginning of the century, with a much smaller population than now, the domestic production of smoked haddock or finnan haddie was approximately 5,000,000 pounds annually. Statistics show a production of 871,000 pounds of finnan haddie and smoked haddock fillets in 1940 (Fiedler 1943). This is partly due to increased demand for fresh haddock, importation of finnan haddie from Canada, and failure of local curers to improve both product and marketing methods. In addition to haddock, varying quantities of cod, cusk, hake, and pollock fillets are smoked, using the same method and producing an article closely resembling haddock fillets. The production of these smoked groundfish fillets amounted to 958,179 pounds in 1940. Some large flounder fillets and fillets of ling cod are given the same cure. Nearly all of these products are known as finnan haddie in the retail market.

The smoking of haddock and related fish is much more important in the British Isles than in the United States, especially in Scotland, where more than half of the "whitefish" (cod, haddock, etc.) cured and smoked. Production of finnan haddie and smoked fillets, together with other smoked whitefish, amounts to more than 40,000,000 pounds annually. A great quantity of haddock is also smoked in England, Norway, Denmark, and Canada. Small amounts of smoked haddock are canned (Jarvis 1944a).

Finnan Haddie—American Method

The finnan haddie is cured almost entirely in New York, Boston and Gloucester, Massachusetts, and Portland, Maine. From these ports it is shipped over the United States. Finnan haddie is also imported, largely from Nova Scotia, in amounts exceeding domestic production. Many customers prefer the imported fish because of its better color, texture, and flavor. Experience is required to prepare good finnan haddie and this applies to all groundfish given a finnan haddie cure. Besides knowing the essential points of the method, practical training is required to smoke a consistently good product.

When haddock are received at the smokery they are already split down the belly to the anal opening and are gutted, in other words, they are dressed as for the fresh-fish trade. The first step in the preparation of finnan haddie is to take off the heads, wash the fish thoroughly and clean out the black membrane lining the belly cavity, using a good stiff brush. The haddock are split down the belly side to within an inch or two of the tail. The knife should not go entirely through the fish. The cut is usually but not invariably made on the right side of the backbone. The fish should now lay out flat in one piece. The next step is to put the cleaned and split fish in a strong brine, testing at least 90° salinometer, for from 30 minutes to two hours. The brining time depends on the flavor demanded by the market for which the product is being cured, the atmospheric temperature and humidity, and the size and condition of the fish.

When sufficiently brined, the fish are taken from the brine and fastened to sticks from which they are suspended in the smokehouse, the napes being stretched out flat and pierced by two nails fixed in the smokesticks. These sticks are about 2 inches square by 4 feet in length, and 3 fish are usually hung from each stick. In some establishments the sticks of fish are hung on the frames and left to dry in the air for a few hours. Other curing plants dry the fish for an hour or two with an electric or blower fan. If the haddock are hung in the smokehouse immediately, a fire should not be started until they have dripped or dried for several hours, or until surplus moisture is not apparent.

The smokehouses are generally like those for curing herring. Some curers are now beginning to use ovens of the type manufactured in the middle west for curing bacon. The haddock are arranged in several tiers, one above the other. The important thing in hanging is to arrange each tier exactly below the one above, leaving enough space between sticks so that the smoke can circulate freely and all the fish will be evenly smoked.

At one time, the fuel used in smoking was oak but since it is scarce, almost any suitable hardwood is used, with rock maple or beech preferred. The smoking usually begins in the evening. A fire is started in the fire pit and allowed to burn from 8 to 12 hours with dampers and ventilators open so that there is not too much smoke. Damp hardwood sawdust is then thrown on the fire so that a dense smoke is created. This is continued for the remainder of the cure. Where the smokehouse is not high, the smoking can be completed in as short a period as 5 hours. The temperature is raised during the second period of smoke cure and kept as high as practicable without cooking or "scalding" the fish. The temperature should average about 80° F. in the second part of the smoke cure. Scalding depends to some extent on the amount of fish in the smokehouse, their moisture content, and distance hung from the fire. If a large amount of fish not well dried is hung in a house, it is much more apt to overheat. Small haddock are given a smoke cure totaling 6 to 7 hours.

When the smoking process is completed, the sticks of finnan haddie are removed to racks in the packing room where they are allowed to cool. If the fires in the smokehouse are extinguished and all doors or ventilators opened, the haddock can be left in the smokehouse until they are cool enough to handle. When cool, the finnan haddie is wrapped in vegetable parchment paper and packed in boxes holding 10 or 20 pounds

Finnan haddie is usually shipped immediately by refrigerated express as it has a very short period of preservation. During warm weather, finnan haddie will keep for only a few days but in winter it may remain in good condition for 2 weeks. As a rule finnan haddie is cured from October to May. For holding over an extended period, the product is kept at a low temperature, usually about 36° F., although one of the largest Pacific coast dealers prefers 20° F., claiming that finnan haddie is safe at that temperature, without damage to quality.

Scottish Method for Finnan Haddie

Smoked haddock or finnan haddie is said to have been originally prepared from dry salt haddock smoked in the fireplace chimneys of Scottish fishermen's cabins. According to one story, the discovery of this product was accidental. This account reports that salt haddock hung from the rafters for storage acquired a smoky flavor, which was relished. Smoked haddock began to be sold as a commercial product, resulting in the development of the present method of cure.

Finnan haddie was first prepared on a commercial scale about the middle of the eighteenth century. At first it was called "findon haddocks," deriving this name from the village of Findon in Scotland, where it was first prepared for market. The name has been shortened to that used at present. A complete description of the Scottish method of curing haddock is given by Duthie (1911). The outline of the Scottish method given here is a summary from that source, together with personal notes taken on a survey of the fish-curing centers of the British Isles.

The haddock may come to the curer in the round, that is, ungutted, or they may have been dressed on board the fishing vessel. If the fish have been previously gutted they have only to be headed when brought to the curing establishment. If not, they must first be eviscerated. The black belly membrane must be removed before washing. The haddock

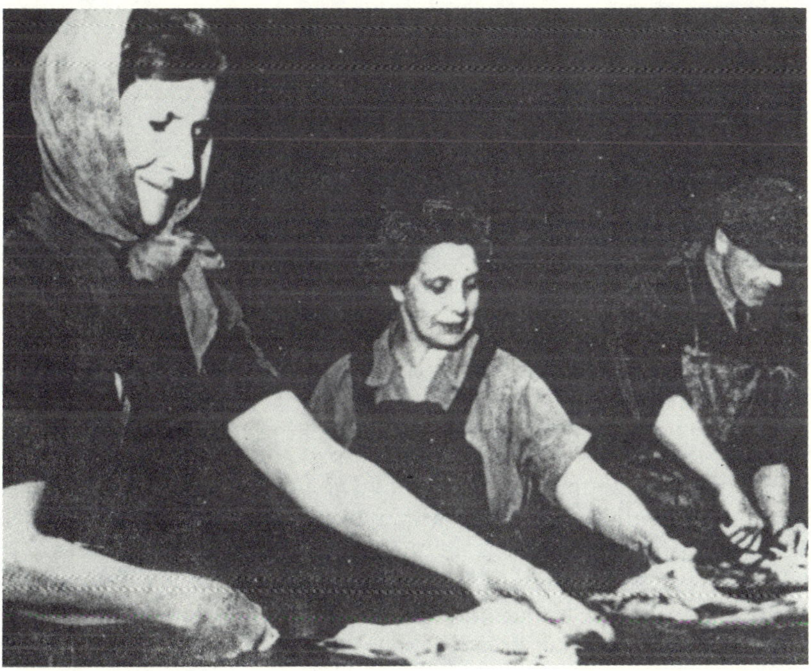

Figure 37. – Splitting haddock for finnan haddie.

Figure 38.—Washing and cleaning haddock for finnan haddie.

are thoroughly washed and freed from all traces of blood, slime, viscera, and "black skin" or belly membrane.

After the fish have been washed, they pass to the splitter (figure 37). The method of splitting is described by Duthie (1911) as follows: They are not split farther down than within an inch of the tail, and a cut is then made on the bone side of the fish. Entering the knife at the shoulder, the splitter makes a deep cut through the small bones to within an inch of the termination of the splitting. The object of this cut is to make the fish look broader. The haddocks as they are split are placed one by one upon a moving belt, two of which generally revolve from the center outwards towards opposite ends of the splitting table. The belt is usually so arranged that each haddock is deposited in a

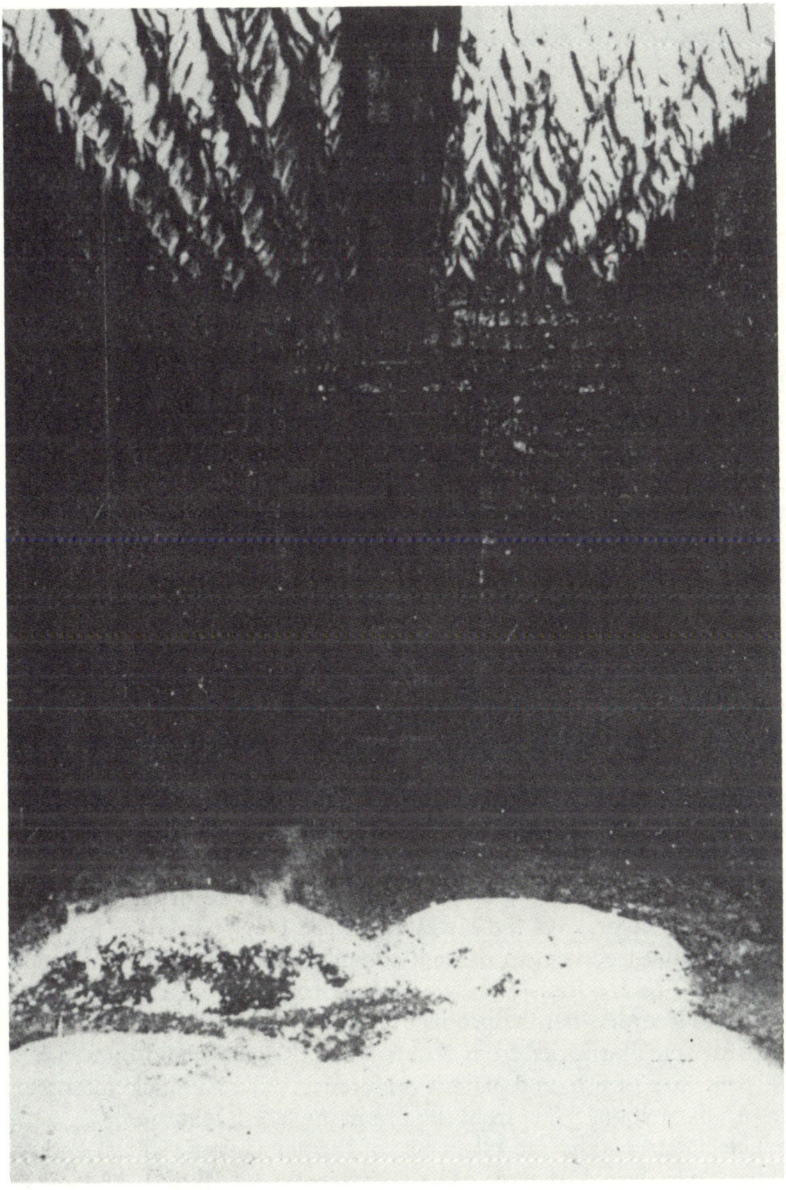

Figure 39. —Interior view of smokehouse, showing the burning of sawdust and chips on fire-pit floor.

chute down which it slides into a receptacle at the end of the washing troughs. Here, a woman takes the fish and brings the blood-bone lightly into contact with a revolving brush, to clean the cavity (fig.38).

After splitting and washing, the haddock are brined in a 90° to 95° salinometer brine, the time required varying from 20 to 30 minutes, according to the size of the fish and market requirements. When they have been sufficiently brined, the haddock are laid upon drippers to drouth, that is, to drain and partially dry before hanging on the smokesticks. The drippers are drying racks, either portable or fixed. The fish are left to drain or dry overnight. In an emergency the fish might be hung immediately after brining, but if this is done they are more apt to fall from the smokesticks than if given a drouthing or drying. At all smokehouses visited in Scotland, the fish were hung on the smokesticks directly from the brine tub and given some draining and drying on the stick before being hung in the smokehouse (fig.39).

A specially constructed smokehouse or finnan kiln is described by Duthie (1911). However, a smokehouse of ordinary type, intended for smoking herring, may be, and often is used for this purpose although haddock must be hung at a lower level than other fish. Practically all smokehouses observed in Scotland curing finnan haddie were of the same type as for kippered herring, sometimes called the "North kiln." The special type apparently is not used so widely as formerly. The lowest tier of fish is hung about 2 feet above the fire in the special finnan kiln described by Duthie. A mixture of peat and softwood sawdust is the fuel used for smoking as a dense cloud of smoke is required to color the haddock. Most of the plants observed by the writer in Scotland did not use peat for fuel but used hardwood or white-wood chips and sawdust. The fish were dyed with annattor before drying to impart the necessary color. As the lowest tiers are hung very near the fire, close watch must be kept to prevent cooking or scalding. The haddock nearest the fire are transferred to the top rows and vice versa during the process in order that the product may be evenly cured. The time required for smoking varies from 5 to 8 hours as determined by weather conditions, size of fish, and market preference. The haddock and other whitefish hang about 8 feet from the fire in the North kiln.

The finnan haddie are taken out of the smokehouse at the end of the cure and allowed to cool. When sufficiently cool they are rinsed in a weak brine solution to remove any sawdust and ashes from the fish. In great Britain, finnan haddies are marketed in boxes holding from 14 to 140 pounds of fish. The most popular size in the trade is said to be the 2-stone box, a container with a capacity of 28 pounds.

Smokies

Haddock too small for curing as finnan haddie, and other small fish, are sometimes made into "smokies" in Scotland. Smokies are a hot-smoked product. No cooking is required. Smokies are also prepared in the Boston area of the United States at a rate of about 30,000 pounds annually. Smokies would have a much wider sale if the public were made acquainted with them by market development advertising.

Small haddock are headed, gutted, and washed in the usual way, but not split. After draining a few minutes they are brined in a 90° to 95° salinometer brine, for 30 to 45 minutes. The brined fish are tied together in pairs by a loop of string twisted around their tails and hung one on each side of a triangular tenter or smoke stick, such as are used in smoking fillets. The sticks of fish are set out to drain off the surplus moisture. About 2 hours will be required for this step, unless the weather is humid and without much breeze, when more time will be necessary.

The smokehouse or kiln may be built in either of two ways. One way is to sink a pit into the ground, lining the wall and bottom with brick. The other kiln is a brick or sheet metal tank built above the ground, with two or three ventilating holes near the ground. The second type is most widely used. The size of the kiln varies according to the requirements. The average dimensions are 12 feet long by 6 feet wide and 4 feet high. A barrel-type smokehouse might be utilized for curing a few fish by this method.

Small chunks of beechwood are the favorite fuel for smoking. A fire is built up over the whole floor of the kiln and allowed to burn brightly, when enough beechwood sawdust is thrown on to give off a dense smoke. The sticks are loaded with fish and set in place across the top of the kiln. A wooden framework cover is then set over the top of the kiln, which is then covered with moist sacks. It is important that the smokehouse be tightly covered, so that smoke and heat cannot escape. Smoking is a rapid process, and the kiln requires little or no attention during the cure. The fish are taken out, practically cooked as well as smoked, in from 25 to 45 minutes, sometimes a little longer if the fish are larger than usual. Smokies are brown-skinned when cured. They are allowed to cool before being packed in small paper-lined wooden boxes. In Scotland the boxes hold one stone (14 lb.) of fish while in Boston the capacity is 10 pounds. Like all hot-smoked products, smokies can be held only a short time.

Smoked Fillets

Smoked fish fillets are fairly well known to the American market where they are often marketed under the name of "boneless finnan haddie." In Great Britain almost any white-fleshed, lean fish, producing fillets of a suitable size, may be cured by this method according to statements made by different fish-curers in interviews with the writer. The method used in England and Scotland is thought worthy of description. It is about the same as that used in the United States and Canada.

The fish are cut into fillets, skinned, and washed. After draining the fillets about 30 minutes, they are placed in a strong brine solution testing about 90° salinometer. The exact time required for brining is said to be determined only by experience and depends on the market for the product. For the British trade it is about a half hour as the average. In the United States fillets are brined from 20 minutes to 2 hours but usually about 1 hour. In Canada, the time varies from 20 to 30 minutes. After draining off the brine, the fillets are dyed. The dye used is annattor, previously mentioned as being used for coloring kippered herring and finnan haddie. The deepness of color given to the fish depends on trade requirements. Some markets demand a fairly lightly colored fish, a shade that might be called straw color while others insist on a deeper shade.

The dye is usually made up in the proportion of one-fourth gallon of dye solution to a bucket of water. In the United States almost any of the colors giving a yellow shade and which are permitted under the Food, Drug, and Cosmetics Act, may be used, except in states where all artificial food colors are forbidden. In some cases, the fish are both brined and dyed in the same operation. If this is done, the dye solution is weaker. Some curers dye the fish before putting them in brine solution. A stronger dye solution is used when the fillets are not brined and dyed in a single operation. They are dipped and withdrawn immediately.

Fresh fillets are not hung on smokesticks like other fish, but hung over triangular wooden tenters or smokesticks, or stretched across a pair of iron rods or arranged on trays with a wire-mesh bottom and wooden frame. They are allowed to drip or drain.

The fire is usually lighted in the evening and the fillets are lightly smoked and dried for a total of 5 to 11 hours. As a rule they are cured over a clear fire with little smoke for 2 or 3 hours, after which they are given a dense smoke for about 3 hours longer. In the United States, the brined and dyed fillets are dried in a current of air, until a thin shiny pellicle or film forms on the surface. This usually requires about 2 hours. The fillets are then hung over three-sided smokesticks, each side being about 3 inches in width, or arranged in wire-mesh trays. The fillets are then cured over a fairly light smoke for 4 hours at a temperature not higher than 90° F. The fillets are then turned so that the side previously resting on the smokestick is uppermost and they are smoked 4 hours longer in a light smoke. The fire is then smothered so that a dense cloud of smoke is obtained, and the fillets are smoked until they are a deep straw color. The fillets are turned once or twice so that both sides will be evenly colored. This operation will take from 4 to 6 hours and the method is used where artificial color is forbidden.

In Great Britain, the fillets, after cooling, are packed in paper lined boxes with average net contents of about a stone (14 lb.) of fish. In America, boxes containing 10 and 20 pounds are used. Smoked fillets must not be overheated or smoked in such a way that they acquire a leathery texture. They should be smoked and salted only to the degree required to obtain the desired flavor, without drying out the moisture contained in the interior. Fillets must be stored at the same temperature as finnan haddie, if they are to be held any length of time.

Chapter Twenty-One

Miscellaneous Fish Smoking

Alewives

Considerable quantities of river herring or alewives were smoked at one time, principally for export to the West Indies. Stevenson estimated the annual production at 5,000,000 pounds in 1899. There was also a local commercial trade in the Delaware and Hudson River areas, and in Maryland, Virginia, and North Carolina. The export trade has disappeared entirely and local commercial curing has decreased greatly. Fishermen and farmers cure quantities of river herring annually for their own use and for sale, but the production is scattered. As much of the production does not appear in the statistical reports, it is difficult to estimate the quantity smoked. It is believed to be about 500,000 pounds. Most of the alewife or river herring smoking is now done in the Chesapeake Bay region of Maryland and Virginia, and in North Carolina. The smoking season is in April, May, and June.

The fresh alewives are washed thoroughly and scaled, as soon as possible after catching. When the washed fish have drained for a few minutes they are put in a tank of 90° salinometer brine. Some salt is scattered over the surface, to strengthen the brine as it becomes diluted. The fish are usually left in brine overnight or for a total of about 12 hours. In the morning the alewives are hung on smokesticks. The stick is inserted through the left gill cover and out the mouth as is done in smoking sea herring. After the alewives are strung on sticks they are dipped in fresh water to rinse them off. The sticks are then placed on racks where there is a good current of air and are allowed to drip and dry until no moisture is apparent on the surface. Drying will require from 2 to 8 hours. The sticks are hung at a minimum distance of 6 to 8 feet from the fire in a smokehouse of the type used for smoking sea herring except that it is much smaller. The alewives are smoked in a dense cold smoke from pine shavings, corncobs or similar material for 48 to 72 hours. The temperature should not be much over 70° F. If the fire is allowed to become too hot, the fish are apt to crack or fall to the ground.

Alewives smoked as described here will keep for about 30 days in the spring and from 10 to 14 days in summer. The shrinkage is not great as the fish are not eviscerated. The curer should obtain about 85 pounds of smoked alewives from 100 pounds of round fish.

A different method is followed in some New England sections. The fish are roused in fresh water but little attention is paid to scaling, as it is claimed that most of the scales will come off in the curing process. The fish are then mixed with salt and the brine is allowed to form naturally. About 25 pounds of salt are used to 100 pounds of fish. The alewives remain in pickle from 3 to 5 days. They are then soaked in fresh water for 4 to 5 hours, and strung on smokesticks, as described for sea herring. After drying for a few hours the alewives are hung in the smokehouse where they are smoked over a low smouldering fire of hardwood and sawdust for 4 to 5 days. When New England alewives were exported to the West Indies, smoking required 2 weeks.

In the Baltimore area and a few other Maryland localities some alewives are smoked as follows: The fresh fish are gibbed like Scotch cured herring, scaled roughly, washed in fresh water and then left for 3 hours in a 90° salinometer brine. When sufficiently brined, they are strung on thin iron rods by piercing the point of the rod through the eye sockets of the fish. The sticks are hung to drain for about 2 hours at the breeziest point around the establishment.

They are then hung in smokehouses made of hogsheads, with a pit below. A fire made of equal quantities of oak and hickory wood is lighted. The fish are allowed to dry over a clear fire for a few minutes. The top is then covered with dampened sacks or old carpet. This is kept moist during the smoking by sprinkling occasionally with water. The fish are cured in steam and smoke for 3 hours, at a temperature of around 170° F. If the temperature goes too high, above 200° F., the fish are apt to drop into the fire. If the fire becomes too hot, it is smothered with hardwood sawdust. The finished product may be eaten without further preparation, as it is cooked as well as smoked. It keeps only a few days after smoking.

Bluefish, Butterfish, Croakers, Flounder, Shad, and Striped Bass

Varying quantities of bluefish, butterfish, croaker, flounder, shad, striped bass, yellow perch, yellow pike, and other miscellaneous species are smoked from time to time. In fact, if the curer is experienced and the proper method is used almost any species of fish may be smoked to make a palatable product. As most of these fish are prepared in small commercial quantities by methods similar to those already described, they are discussed together in the following section. The butterfish (*Poronotus triacanthus*) is a small food fish with a deep body, so that it looks almost round. Because of its shape and bright silvery color, it is given the secondary common names of "dollarfish" and "pumpkinseed." The butterfish is taken commercially mostly in the middle-Atlantic states. It is caught largely in traps and pound nets, but also to some extent by seines. Statistics show a production of 480,700 pounds of smoked butterfish in 1940. The butterfish are washed but not gutted. They are brined from 2 to 6 hours in a 90° salinometer salt solution. The fish are then strung on smoking rods, usually run through the eye sockets. After draining and drying for a few hours, the butterfish are hung in the smokehouse where they are cured in a cool smoke for 4 to 5 hours. When the fish have the desired amount of color, the fire is built up and they are hot smoked at around 200° F. for about one hour.

The croaker (*Micropogon undulatus*) is one of the common marine food fish from the Middle Atlantic states to the Gulf of Mexico. It averages about 10 inches in length and has an average weight of 1/2 to 1 1/2 pounds, depending on locality. The maximum weight is from 3 to 4 pounds. The croaker is taken mostly in pound nets, haul seines, and otter trawls, and to some extent by hand lines and miscellaneous gear. It has a lean, white, firm flesh.

The larger croakers are smoked occasionally. The head is removed, retaining the collar bones. A cut is made down the belly to the vent and all viscera or other offal are removed from the body cavity which is scraped and cleaned thoroughly. Then the fish are split down the belly side to within an inch or two of the tail. The knife should not go all the way through the fish. After the croakers are cleaned they are washed, then placed in a strong brine from 30 minutes to one hour. The brined fish are fastened to smokesticks as in curing finnan haddie, so as to lay out flat in one piece. They are dried for a few hours. The drying process is shortened and standardized by using an electric or blower fan. The sticks of fish are then hung in a smokehouse and given a cool smoke of light density from 2 to 4 hours. The fire is then covered with sawdust and the croakers are held about 2 hours longer in a dense smoke. The temperature should not be higher than 100° F.

The flounders (*Pleuronectidae*) include a variety of species of flatfish. These are fish with the body much compressed, both eyes on one side of the head, with the blind side colorless and usually the underside in the water. The two most important commercial species on the American coast are the winter flounder and the summer flounder. The average weight is from 1/2 to 3 pounds, but they may weigh up to 15 pounds. They are caught mostly by otter-trawls, line-trawls, and hand-line. Small flounders, those weighing about 1/2 pound, are gutted, washed, and brined in a 90° salinometer salt solution for 2 hours. They are then strung on smokesticks, drained, and air-dried for a few hours. They are cold smoked for from 8 to 10 hours. Some smokers then build up the fire and give the flounders an additional hot smoking for about 30 minutes. Large flounders are cut into fillets which are smoked in the same way as cod and haddock fillets.

The shad (*Alosa sapidissima*) is an important food fish on both Atlantic and Pacific coasts, but the largest fisheries are in the Middle Atlantic and Chesapeake Bay areas. The average weight is 3 pounds for buck shad and 4 3/4 pounds for roe shad, with a maximum of 15 pounds for the species. Shad are caught mostly in pound nets, traps, and seines. Some shad are smoked in the Chesapeake Bay region. The method is much the same as for croakers. Some smokers dry and color the fish by cold smoking, then hot smoke and cook them for about an hour. This is also the method usually followed in smoking the bluefish (Pomatomus saltatrix). The method used for smoking shad on the Pacific coast is much the same as for salmon. Some shad are lightly smoked or kippered and canned on the Columbia River. This is described in detail elsewhere (Jarvis 1944a).

The striped bass (*Roccus saxatilis*) is a favorite food fish of the Atlantic coast. It has also been introduced to the Pacific where it is now quite common especially in the San Francisco Bay area. It is especially abundant from Massachusetts to North Carolina. The usual weight is from 3 to 5 pounds. Striped bass are taken in traps, pound nets, gill nets, seines, and by hook and line.

The striped bass is occasionally smoked with good results, using the finnan haddie cure. It has also been smoked like salmon after first curing in a salt-sugar-spice mixture.

The Florida sailfish (*Istiophorus volador*) is usually regarded as a sport fish and is taken by trolling. Some of the catch brought in to ports in southern Florida is smoked, using the method as outlined for sturgeon and catfish.

Carp

Carp (*Cyprinus carpio*) is one of the less important smoked products. It is an under-utilized fresh-water species which makes a smoked product of good quality and could be prepared in much greater quantity, especially in the North Central states, if some effort was devoted market development. Production is apparently increasing slowly. Statistics indicate a production of 106,000 pounds in 1930, 116,000 pounds in 1935, and 173,000 pounds in 1940.

There is considerable variation in commercial methods of smoking carp but the following description may be considered typical: The fish are headed, gutted and cleaned thoroughly, then cut in steaks or chunks weighing from 1/2 to 1 pound each. The steaks are brined for 10 to 12 hours in a brine testing from 60° to 70° salinometer. The length of time varies with the size and condition of the pieces. Fish not in best condition are given a heavier brining for a longer period.

In some instances the pieces are hung on smokesticks; in others, they are placed on wire-mesh-bottomed trays, previously oiled to prevent the fish from sticking. Some smokers then sprinkle the fish with spices and condiments, such as red pepper, poppy seeds, cloves, or bay leaves.

The chunks of carp are then placed in a smokehouse for a period averaging 3 hours and cooked or "dried" over a hot fire without much smoke. The temperature in the smokehouse during this period is about 225° F. The fire is then broken down and dampened with sawdust to build up a dense smoke. The carp pieces are then smoked for about an hour at a temperature ranging from 90° to 110° F., which should give sufficient color and flavor. When the fish has cooled, the pieces are wrapped in parchment paper and packed in boxes.

Catfish and Paddlefish

Other fish are sometimes smoked as a substitute for sturgeon. Two species are so cured, the channel catfish (*Ictalurus lacustris*) and the paddlefish or spoonbill catfish (*Polyodon spathula*). The annual production is about 10,000 pounds of the former and 150,000 pounds of the latter species. The fish are taken in the lower Mississippi River, and most of the smoking is done at St. Louis and Chicago. Most are taken by set lines, though some are caught in haul seines.

Since catfish were first smoked as a substitute for smoked sturgeon they are cured by the same method as sturgeon. The fish usually come to the smokehouse from the fresh-fish dealers already gutted and without heads. As the first step in preparation, the fish are skinned and cut into chunks weighing about 1 or 1 1/2 pounds each. The pieces are washed, then brined or pickled in tubs for 6 to 8 hours. Brining may be done in either of two ways: The pieces may be rubbed with salt and packed in a tub with more salt scattered over them, where they are allowed to form their own brine, or they may be placed in a strong brine (about 90°) salinometer) at once. When the pieces have been sufficiently brined they are rinsed in fresh water to remove surplus salt or other foreign materials. Then they are strung on smokesticks or placed in wire-mesh bottomed trays, after which they are air-dried for 2 hours. The fish are then placed in the smokehouse to be smoked for 8 to 9 hours in the same way that sturgeon is smoked. One hundred pounds of dressed catfish yield 65 to 70 pounds of the smoked product.

At various points in the Mississippi Valley and Great Lakes areas, small catfish (*Ameiurus* sp.) are smoked whole, like lake herring. They are split to the vent and eviscerated. The head and, in some instances, the skin, are left on. The fish are mixed with salt and left in the tubs to brine. When sufficiently struck, they are smoked for a few hours following the procedure described for lake herring. Most of the small catfish smoked are prepared by individuals for home use, and only small amounts are cured commercially.

Eels

Smoked eel has always been considered one of the most important smoked-fish products in Germany. Large amounts of eels were imported into Germany from Denmark and other European countries, and even from Egypt. Smoked eel is also a favorite smoked-fish product in the Netherlands. The smoking of eels is said to have been introduced into the United States by early Dutch colonists, but the greatest demand in this country has come at a comparatively late date from the people of German and other North European stock. In the United States, eels are smoked principally in New York, Philadelphia, Chicago, and Milwaukee. The production varies between 100,000 and

300,000 pounds annually. The amount smoked was 225,000 pounds in 1940.

The common eel (*Anguilla bostoniensis*) is widely distributed, being found along the Atlantic coast from Canada to Key West. It may be distinguished by the slender snake-like form of body, the low vertical fins, devoid of spines, and the absence of ventral fins. The common or freshwater eel varies in length from 2 to 3 1/2 feet and reaches a maximum weight of 16 1/2 pounds.

Eels are caught in eel pots, which are small cylindrical traps, usually made of heavy galvanized screen with a 1/4 or 1/2-inch mesh, and funnels. The traps are anchored on the bottom of a stream moored to a line or with a float to mark the location. Eels are also caught by spearing. This is usually done at night. The fisherman carries a torch or lantern, which attracts and dazes the fish, making it comparatively easy to spear. Eels are taken to some extent with handlines, trawl lines, fyke nets, pound nets, haul seines or drag nets, otter trawls, and floating traps.

Some eels are received fresh but probably the greatest quantity come to the smokehouse frozen. If frozen, they must be defrosted by soaking in a tub of cold water for 4 to 10 hours, depending on size. The eels are dressed after the surface slime is removed by scraping or rubbing with fine salt. They are split down the belly from the head past the vent and eviscerated. If possible, the large vein along the backbone should be pulled out or cut out. The larger eels are cut into cross-sections from 1 to 2 inches thick.

The eels may be given the proper amount of salt cure by rubbing them in a tub and allowing the brine to form naturally, or they may be placed in 100° salinometer brine. As with other smoked products, the length of time in brine depends on the weather, method of salting, length of preservation desired, and local market preference. Eels may be in brine from 2 to 24 hours.

Dry salting requires less time than brine salting. A German reference gives the time as 5 or 6 hours (Hoffman: undated). Tressler (1923) reports it is about 24 hours. Stevenson (1899) states that about 7 hours are used in the Great Lakes areas, 2 hours in New York.

On removal from the brine, the eels are rinsed in fresh water to remove any surplus salt or slime. Sometimes they are scrubbed with a stiff-bristle brush. The fish are next strung on iron rods about 1/4-inch thick, the rod passing through the head of each eel, or through the throat cartilage and out of the mouth. In other establishments, the eels may be pierced through the head, twisted around the rod, and pierced again through the body near the tail. If the eels are large and have been cut into sections, the pieces are arranged on wire-mesh-bottom trays, which have been oiled to prevent sticking. Some spines such as caraway, peppers, bay leaf, coriander or cardamom seed, or a combination of several of these, may be scattered on top of the slices. If possible, the eels should be dried in the open air for a few hours but if this is not practical they may be dried off by fans.

There is also great variation in the smoking procedure. Some eels are smoked for 2 or 3 hours over a low, clear fire without much smoke. Then a hot fire is built up, covered with sawdust to make a dense cloud of smoke and the eels are hot-smoked and partly cooked for 2 hours longer. Other curers hot-smoke the eels over a hot fire of corn cobs until they are nearly cooked; then the fire is partially smothered with sawdust which produces a dense smoke that colors and partially dries the eels. In some Middle West smokehouses, smoking is done at an even temperature near 100° F. for 6 to 8 hours. In one or two eastern centers, the eels are given a mild smoke for 4 or 5 hours until they have acquired the proper color, when the fire is built up and the eels are hot-smoked or cooked for 30 minutes at about 240° F.

One of the methods for smoking eels followed in Northern Europe, especially Germany, is somewhat different from methods described previously. The head, skin, tail, and viscera are removed, and the eel is split lengthwise into two fillets, removing the backbone, with many of the small rib bones attached. The fillets are then covered with a strong brine (about 90° salinometer) where they are left for 6 hours. They are then wiped dry with a clean towel and covered with the following preparation, which has been pounded into a paste in a mortar: one large anchovy, 1 ounce of fine salt, 8 ounces of powdered sugar, 1 ounce of saltpeter, and sufficient butter to make a smooth paste of the ingredients. The eel fillets are coated with this preparation, then rolled up tightly in the form of a cylinder, beginning with the tail end. The cylinders are tied in the middle with a cord to hold them together. They are then sewed up in a cloth which covers the sides and allows the end to project. These cylinders are hung in an ordinary smokehouse where they are cured in a cool, dense smoke for 5 to 6 days. After cooling they are ready for market.

Eels are smoked and marinated in Italy and Dalmatia by the following method: The eels are thoroughly cleaned, headed, and skinned. Larger eels are cut in slices. The eels are then fixed on spits, or if sliced, arranged on trays, and smoked over a hot fire of oak and ash wood for 40 minutes at a temperature of about 225° F. When the eels are cooled they are packed in kegs, holding from 25 kilograms (55 pounds) to 55 kilograms (121 lb.). Spices, principally bay leaves, whole peppers and cloves, are scattered between the layers of fish. When the container is full, a solution of 6 percent vinegar and 3 percent salt is poured in until the keg will hold no more. This preparation is known as "Comacchio" type eels (Jacobsen 1926).

Goosefish

The goosefish (*Lophius piscatorius*) is a large, sluggish fish found on the North Atlantic coast from Nova Scotia to Cape Hatteras. Local names are anglerfish, monkfish, and allmouth. It reaches a length of 4 feet and a weight of 40 pounds. Although palatable and a good food fish, most of these fish are thrown back after removal from the nets. Statistics give the catch as 104,000 pounds in 1943. The goosefish or angler has not heretofore been prepared as a commercial smoked-fish product. A method has been developed by the U.S. Fish and Wildlife Service. As a result, hot-smoked angler is now being sold on a small scale in the New York market. It is an appetizing article which should find a wider sale, if aided by market development work.

The fish are landed, headed, and eviscerated. The first operation at the curing plant is to wash the fish thoroughly in fresh water to remove slime, after which they are skinned. The flesh is removed in two fillets by cutting along both sides of the backbone from head to tail. The fillets are cut into portions which weigh approximately 1 pound each after smoking. These portions are placed in a 50° salinometer brine where they are left from 16 to 20 hours. The length of time depends on the thickness of the pieces, the preference of the trade for which the fish is being cured, and weather conditions. After salting, the pieces are freshened in cold running water for 30 to 60 minutes and are spread on wire-mesh-bottomed trays brushed with edible oil to prevent the fish from sticking. Pepper, garlic salt, or other condiments to suit the taste are sometimes sprinkled on the fish before drying.

The fish are dried in a current of air at about 70° F. until all surface moisture has disappeared and a thin skin or pellicle has formed. The trays are stacked in the smokehouse. The fish are cooked for 4 hours, the temperature being raised slowly to avoid case-hardening, and then maintained at 170° to 170° F. A heavy smoke is then started, using oak sawdust for fuel and the cure is continued for another 1 1/2 hours at the same temperature. The fish should then have the desired flavor, color, and texture. The doors of the smokehouse are opened immediately and when the fish is cool enough to handle the chunks are wrapped in cellophane or vegetable parchment and stored under refrigeration until sold. Hot-smoked anglerfish is similar to other hot-smoked fish products in appearance and uses.

Halibut

Smoked Atlantic halibut (*Hippoglossus hippoglossus*), today a product of minor importance, is mostly cured on order for a limited market. At one time it was an important cured fishery product. According to Stevenson (1899), 3,000,000 pounds of halibut were smoked in the United States in 1872, and the annual production averaged 2,000,000 pounds in the latter part of the 19th century. The annual output today probably does not amount to more than 15,000 pounds.

There are several reasons for the great decline in the preparation of smoked halibut: First, the depletion of the Atlantic halibut fishery, now at only a small fraction of its former level; second, long voyages are no longer made to Davis Straits and Greenland to salt halibut, which was the source of raw material for most of the smoked halibut; third, the development of freezing has made available a better method of preservation; and fourth, the price of fresh halibut has been so high that curers can rarely afford to smoke it. Most of the halibut is taken by trawl line (long line); some is taken by hand line and otter trawl incidental to other fisheries.

As salted halibut is not ordinarily available today, smoked halibut usually is prepared from fresh fish. As the first step, the halibut is cut into "fletches" or strips of boneless flesh, with each side cut into two fletches. The fletches are thoroughly washed, drained for a short time, then salted down in butts like those used in curing codfish. The strips are rolled in salt and packed in like codfish, skin side down, except for the top layer, with a scattering of salt between each layer of fish. The brine is allowed to form naturally. If the amount of brine is insufficient, 100° salinometer (saturated) brine may be added. After one or two weeks, the fletches may be taken out and kenched, or may be allowed to remain in the butts until ready for smoking. Approximately 10 days are required for the salt to strike through or salt cure the flesh.

When the halibut are to be smoked the fletches are taken out of salt, washed thoroughly, and scrubbed with bristle brushes to remove encrusted salt. They are then placed to soak in fresh water for 3 or 4 hours. The water is then changed and the halibut are again soaked for about the same length of time. Soaking removes excess salt and softens the flesh for smoking. At the end of the soaking period the fletches are water horsed for 5 or 6 hours, that is, piled in stacks or kenches, skin side up, and weighted down heavily to press out moisture. After water horsing, the fletches of halibut are spread out on flakes like those used for drying codfish. Here they are sun-dried for a total of 24 hours with the drying time spread out over 3 or 4 days, or even longer if the weather is unfavorable. The fish are placed in small piles and covered with flake boxes during the night or in rainy weather. After drying, the fletches are cut in small pieces, from 2 to 6 pieces to the fletch, with a gash in each piece where the flesh is thin and the skin appears tough.

The fletches are then usually strung on smooth, round, hardwood sticks about 2 feet long and 3/4-inch in diameter, though sometimes rods of thick steel wire 3 1/2 feet long are used. The rods are passed through the gashes in the chunks of halibut. From 5 to 7 pieces are hung on sticks 2 feet long, and from 8 to 12 on 3 1/2-foot rods. The pieces are spaced 2 or 3 inches apart so that the smoke will penetrate and color evenly. The sticks when loaded are passed into the smokehouse, which is of the standard type, such as is used in smoking salmon or herring. Practically all of the halibut smoking is done in Gloucester, Mass.

Oak chips or sawdust are preferred as fuel for smoking. Chips are generally used in starting the fires, with sawdust to smother the flames. Fires are built along the two long side walls of the fire pit, leaving the middle empty. They are kept going until the smoke cure is complete, which requires from 2 to 5 days, depending on the weather and the size of the pieces. Doors and ventilators are left slightly open so that there will be a constant circulation of air since for a good cure halibut must be kept cool and as dry as possible. It also takes the smoke better under this arrangement. The loss in weight is about 70 percent. A halibut weighing 100 pounds round will yield 30 pounds of the finished product. A good smoked halibut should have a straw-yellow color. Most of the halibut now smoked are cut in small strips and packed in glass tumblers to a net weight of about 5 ounces. Metal tops are vacuum sealed on the containers, which are not heat processed.

Kingfish and Spanish Mackerel

The kingfish, or king mackerel (*Scomberomorus cavella*) also known as the cero or painted mackerel, is one of the choicest food fish of the South Atlantic and Gulf Coast regions. It is sometimes smoked as far north as Morehead City, N. C. Regular commercial production is confined to Florida where it is cured by fishermen for local sale. Kingfish are taken mostly by trolling lines, though some catches are made by runaround gill nets, haul seines, and hand lines.

As soon as possible after landing the kingfish are cut into 2 sides, removing the backbone. The sides are washed thoroughly, drained for a few minutes, then placed for about one hour in a 40° salinometer brine which is chilled with ice. The object of this step is to leach diffused blood from the flesh, and to make the flesh firmer for handling during curing. The brined kingfish are drained for a few minutes, then dredged in packers fine salt in a shallow box. Some salt is rubbed gently into the flesh. The sides are picked up with as much salt as will cling, then laid in a box. They are left in salt from one to two hours, depending on the size of the fish, and the weather. If the sides are large and the weather humid, the longer salting is followed.

When it is judged that the fish has enough salt the sides are washed in fresh water, to remove all traces of salt, then fixed on hangers and dried in a current of air, either from a fan or in a good breeze. The kingfish are dried until no surface moisture is apparent, and preferably until a thin shiny skin has formed on the surface. The sides are then hung in the smokehouse, from 4 to 6 feet above the ground. A low fire of dry hardwood is lighted with the smokehouse ventilators open, and the fish is cured without much smoke for 24 to 36 hours. The ventilators are then closed, the fire is smothered with hardwood chips, sawdust or bark, and a dense smoke is built up. The kingfish are cured in this smoke from 6 to 12 hours, depending on the color and flavor preferred. Some consumers prefer a lighter color and flavor, others a glossy brown color and pronounced smoke flavor. Some curers brush the flesh surface of the sides with vegetable oil just before making the dense smoke. This is to help in giving the desired glossy appearance. The temperature of the smokehouse should be kept below 90° F. After cooling, the sides are wrapped in oiled or parchment paper and packed in light wooden boxes. Smoked kingfish are cut into thin slices like smoked salmon, and the slices are sometimes grilled like ham or bacon.

The above method is also used in curing Spanish mackerel (*Scomberomorus maculatus*). If mackerel are being cured, the salting period is from 30 to 60 minutes, the initial period of the smoke cure from 8 to 12 hours, and the final portion in dense smoke from 2 to 4 hours.

Spanish mackerel and occasionally kingfish are also hot smoked, giving a product that is preferred by many consumers as it requires no additional cooking or preparation. The Mackerel are split into sides. Sometimes, especially with larger fish, the backbone is removed. The fish are washed and soaked to remove blood, as described above. A brine is then prepared in the following proportions: 2 pounds salt; 1 pound sugar; 1/2 ounce saltpeter; 1 ounce crushed whole black peppers; 1 ounce crushed cardamom seeds; and 1/2 ounce crushed bay leaves.

The ingredients are made into a brine that will test about 90° salinometer, or by the more usual local method, a brine that will float a potato. The amount of ingredients is increased according to the quantity of fish cured. Various spice mixtures are used, changing variety and quantity according to personal preference.

The mackerel are left in this brine for 2 to 4 hours depending on the size and thickness of the fish, the amount of fat, and individual preference. Some consumers require a less salty taste than others. The exact length of time must be determined by experiment. After rinsing off the fish in fresh water, they are placed on hangers or wire-mesh-bottom trays which are set outside in a cool, shady, breezy place to dry for about 3 hours before hanging in the smokehouse, at least 6 feet above the fire.

For the first 8 hours that the mackerel are in the smokehouse, they are cured in a dense cloud of smoke at as low a temperature as possible, not more than 90° F. The fire is then built up until the temperature is between 130° and 150° F., with less smoke. The fish are cured at this temperature for about 2 hours, or until the flesh has a glossy light brown color, partially cooking or hot-smoking the mackerel. The fish are allowed to cool before handling. Any moisture on the surface is then wiped off, the fish are packed in waxed paper or vegetable parchment paper and are ready for use. Unless held in refrigeration the period of preservation is brief, averaging about 10 days. With a storage at 40° F. or less, smoked mackerel can be kept for an average of 60 days.

In some cases the mackerel are brushed over lightly with vegetable oil, usually cottonseed, either just after finishing the cold smoking part of the process, or on taking them out to cool after smoking. Another method of handling this fish after smoking is to cut the flesh into fingers the length of a No. 2 can or pint jar. After skinning, these pieces are packed asparagus-style into the container. Vegetable oil is then added until the spaces between the pieces of fish are filled and there is a layer of oil up to within a quarter-inch of the top. The containers are sealed and stored in a cool place until used. As this product is not sterilized it will not keep indefinitely and must not be held at room temperature.

Mackerel

Mackerel (*Scomber scombrus*), also known as Atlantic or Boston mackerel, is smoked in the vicinity of Boston and New York. Figures on production are not available as this item is included under miscellaneous smoked products in U. S. fishery statistics. It is estimated that about 30,000 pounds of mackerel are smoked annually. A small amount of smoked mackerel is canned. Most of the mackerel are caught by purse seines, though some is taken by pound nets, floating traps, and miscellaneous types of gear.

The method generally used in smoking mackerel is much the same as for kippered herring. The mackerel must be fresh and firm. "Belly-blown" fish do not make a good product. The mackerel are delivered from the wharf to dressing tables where the fish are split down the left side of the backbone from the tip of the head almost to the tail, or as for salting, and opened to lay flat in one piece. Gills, viscera, blood, and membranes are cleaned away. The fish are washed well, then soaked in fresh water from 15 to 30 minutes to remove diffused blood from the flesh. After draining a few minutes the mackerel are brined for about an hour in a brine testing about 80° salinometer. The proportion of brine to fish is usually two to one.

When sufficiently brined, the fish are fixed on smokesticks of the same type used for kippered herring. The heads are impaled on nail points on each side of the stick, with no two fish touching. The mackerel are allowed to dry for several hours until moisture is not apparent on the surface. Fans or blowers may be used for drying. The sticks or mackerel are then hung at least 8 feet above the fire and smoked for about 8 hours in a light cool smoke. The temperature should be less than 80° F. The fire is usually of hardwood chips, built on the two long sides of the smokehouse with an empty space in the middle. The ventilators may be left partly open as a dense smoke is not desired. When the smoked mackerel are cool, the individual fish are wrapped in vegetable parchment paper and packed in light wooden boxes holding about 10 pounds.

The process described by Stevenson (1899) is still used occasionally in New York. The fresh mackerel are washed, but not dressed, being brined in the round. They are left in a 90° salinometer brine for 12 to 14 hours, then removed and opened at the vent with the point of a knife to drain out the brine in the belly cavity. The mackerel are next fixed on smokesticks, with the stick inserted through the left gill cavity and out of the mouth.

After air drying for 2 or 3 hours they are hung in the smokehouse, where they are given a cool smoke for 4 or 5 hours until properly colored. Then the fires are built up to a temperature between 150° and 200° F. and the fish are given a hot smoke for from 1 to 2 hours.

The Fish and Wildlife Service has made a study of the preparation of smoked mackerel (Stansby and Griffiths 1943) and has devised a method for curing smoked mackerel of better quality than that usually found in the market.

The method is briefly as follows: The mackerel are dressed, removing the heads, and splitting down the backbone to lay open in one piece. Viscera, black skin, and other offal are cleaned out of the body cavity. The dressed fish are soaked in fresh water for 30 minutes to remove diffused blood from the flesh. They are then brined for 15 minutes in a 10% (38.5° salinometer) salt solution, used in the proportion of 2 pounds of brine to 1 pound of dressed fish. After brining, the fish are hung overnight in the smokehouse in a current of air created by means of a blower. This forms a thin film or pellicle on the flesh, resulting in a smooth, glossy surface, which absorbs and is colored uniformly by the smoke.

For a more rapid cure, forced drying may be used. The fish are drained for a short time, then hung in the smokehouse, which is heated to 100° F. and a current of air is drawn through the house by means of a blower. Drying is completed in about 4 hours.

A smokehouse in which temperature, humidity, and smoke density may be controlled, is used in smoking this product. (Lemon 1932). The fish are smoked at a temperature of 100° F. and at as low a relative humidity as possible. Air-dried oak sawdust is used to produce a smoke of moderate density. Best results are secured by cold-smoking for 3 to 4 hours. The smoked mackerel then has a mild smoke flavor which does not hide the natural flavor of the fish. If a strong smoke flavor is desired the mackerel may be cold-smoked for 8 hours.

Mullet

The striped mullet (*Mugil cephalus*) is smoked commercially in small quantities on both coasts of Florida and as a noncommercial article for home consumption in all of the South Atlantic and Gulf Coast states. The commercial production of smoked mullet amounts to about 35,000 pounds annually. Some data have been collected on noncommercial mullet smoking but they are incomplete and unsatisfactory. It is estimated that production of noncommercial smoked mullet amounts to about 70,000 pounds annually. The methods of catching and handling mullet have been outlined under the discussion of dry salt mullet.

There is no regularity of procedure in smoking mullet and the product is variable in quality. Some mullet are hot-smoked by the method given for smoking Spanish mackerel. The method described here is recommended as giving the best results.

The mullet are split along the back just above the backbone, almost to the tail, so they will lay flat in one piece. The belly portion is left solid. The backbone is left in. All traces of viscera, membranes, black skin, and blood are cleaned out, taking special care to remove the coagulated blood and the kidney. The head may or may not be removed, depending on individual preference of the smoker. It is often removed when smoking the larger mullet. The heads are hard, and are difficult to cut mackerel style and to clean of waste material. They have no edible material, but since the smoked fish sell by weight, some curers like to retain the heads, thus reducing the loss of weight in curing. If the head is cut off, the hard bony plate just below the gills should be allowed to remain as it will be needed to carry the weight when the fish are hung on rods for smoking. If the "collar bone" is cut off, the mullet may pull loose and drop from the sticks.

After splitting, cleaning, and washing, the mullet are dropped into a salt brine testing about 40° salinometer. The proportion should be about 1 part salt to 10 parts water. The fish are left in this brine for 30 minutes to soak out the blood diffused through the flesh. At the end of this time they are taken from the brine, rinsed, and freed of remaining traces of blood or other offal. After draining for a few minutes the mullet are dropped into a shallow box, partly filled with packers fine salt, where they are dredged about in the salt. When the fish have acquired a coating they are picked up with as much salt as will cling to the body and packed in even layers in a tub or box.

The mullet are left in salt from 1 to 3 hours (occasionally up to 8 hours) depending on weather, size of the fish, fatness, and length of time for which preservation is desired. The exact length of time must be determined by the smoker according to varying local conditions. When the fish are taken out of salt they are rinsed in brine and all visible traces of salt or dirt are scrubbed off. The mullet are then laid on chicken-wire drying racks on legs 3 feet high, kept out of the direct rays of the sun but located where a good breeze can reach them. Wire drying racks are desirable so the fish can dry on both sides; if laid on boards one side will remain wet. Before being hung in the smokehouse the mullet are given an average of about 3 hours drying, or until a thin film is formed on the surface. If hung shortly after being taken out of salt, with the surface still moist, the fish require longer smoking, do not color and dry so well, and the appearance of the surface is not so attractive.

The mullet may be placed in the smokehouse on wire-mesh trays which are much easier to handle than by hanging on sticks with rows of nail points, or on thin iron rods. In no case should any two fish touch as this obstructs the drying and coloring action of the smoke. If hung on iron rods, more fish mat be smoked at one time and will be better smoked, with a clearer color. Pointed iron rods are run through the mullet just below the hard, bony plate at the neck, one rod on each side (fig.31). Thus each fish hangs from two rods. From 12 to 14 fish can be hung on a set of 2 rods 3 feet long.

The fire is started an hour or two before the mullet are hung in the smokehouse. Almost any hardwood or nonresinous wood may be used for fuel. Pine or other pitchy woods give the mullet a bitter taste. In experimental work it was noted that while buttonwood and tupelo were satisfactory over a short curing period, they gave mullet an unpleasant flavor is used for any length of time (Jarvis 1933).

Some woods used in the southern States for smoking mullet are scrub oak, live oak, hickory, sweet bay, river mangrove, palmetto roots, and coconut husks. Sweet bay is considered to give mullet the best flavor. The odor is also faintly aromatic. Orange wood gives a particularly pleasing flavor and a more attractive color than some of the others. The fires should not give off too much smoke during the first 8 to 12 hours of the process; after that a dense cloud of smoke is built up for the balance of the cure. The temperature in the smokehouse should not be allowed to fluctuate. In experimental smoking of mullet a fire made from two short chunks of wood about 2 feet long and the thickness of a man's arm, was usually found to be sufficient. The standard test for temperature is to put the hand into the smokehouse. The air should not feel distinctly warm on the hand. The mullet are smoked for 20 to 24 hours if they are to be kept for about 10 days and from 3 to 5 days if it is wished to keep them for some time.

After the mullet are taken out of the smokehouse they are dried for an hour or two in the air, then a little fine salt is sprinkled on each fish. They are wrapped in paper, packed in wooden boxes, and stored in a dry cool place.

Good results have also been obtained in experimental smoking of mullet with a hot-smoked product. The method is identical with the foregoing one up to and including 8 hours of cure in a light cold smoke. A dense smoke is then built up with a temperature of 200° to 240° F., and the fish are hot smoked for 30 minutes, if small; up to 60 minutes for medium or large-sized mullet. The hot-smoked mullet have only a short period of preservation, about 10 days at a maximum. They are sold principally at roadside stands for immediate consumption.

Sablefish

Barbecued sablefish (*Anaplopoma fimbria*), also known as kippered black cod, smoked black cod, and barbecued Alaska cod, are cured and used in a few localities on the Pacific coast; in the vicinity of Chicago and Milwaukee, in the Great Lakes district, and in New York and Philadelphia on the Atlantic coast. The sablefish has firm, white, flaky flesh, and a full, rich flavor which make a smoked product of excellent quality. Although smoked sablefish is pronounced delicious by the epicures, it has not found favor with the general public except in a few local markets. Barbecued sablefish should be cured much more extensively. If progressive fish-curers would promote the production and marketing of smoked sablefish they would create a steady demand for a profitable article. The United States produced a total of 1,175,000 pounds of smoked or kippered sablefish in 1940.

The largest part of the sablefish catch is made by long lines, although with the expansion of trawling on the Pacific coast, increasing quantities are now being taken in otter trawls. Sablefish are taken incidental to the halibut catch, but as the halibut are more valuable and the boats are restricted to a certain number of trips, the fishermen bring in the higher priced fish, avoiding and at times even discarding the lower priced sablefish. The development of trawling may change this to some extent.

Sablefish or black cod are almost too fat to cure by cold smoking, although a small amount is prepared by that method. The rich, flaky flesh is much more suitable for hot smoking (kippering or barbecuing). Fresh or iced fish are sometimes used as a raw material but as in kippering salmon, frozen fish are preferred, and for the same reasons. The flesh is firmer and therefore easier to prepare for smoking, and the curer is not dependent on a fluctuating seasonal catch but has an assured supply of raw material to be drawn on as required.

There is some variation in the method of handling. The sablefish are received by the smoker frozen, headless, and gutted. The frozen sablefish are first thawed in a tank of cold water. The fish are left in a tank overnight or for an average of about 12 hours. Some smokers use running water as it reduces the time required for thawing and because they have been found that in a tank of still water the fish would freeze together in a mass unless stirred vigorously from time to time. The sablefish are cleaned and dressed while still partly frozen. The body may or may not be scaled, the fins are removed and the belly cavity is cleaned out. The fish are then split into sides. The splitter first makes a short cut just under the backbone and above the anal fin, in the region of the vent. A second short cut is then made at the same place, just above the backbone. A short preliminary cut is made just above the end of the backbone which can then be removed with another stroke of the splitting knife. The thin, belly flesh on each side is trimmed away. The belly trimmings may be smoked as strips or salted in barrels. The trimmed sides are cut into chunks which will weigh approximately 1 pound each after smoking. Pacific coast smokers of sablefish now often split sablefish in two sides and cut them in chunks without removing the backbone. This is guess work, of course, but experience brings remarkable accuracy. The average piece is approximately 6 by 3 inches.

The pieces of sablefish are placed in a tank to cure in a 90° salinometer brine for 2 1/2 to 3 hours; the general rule being to brine this fish one-half hour longer than the salmon under the same conditions. After brining, the fish are drained 5 to 10 minutes to remove surplus moisture. On the Pacific coast, the chunks are placed in wire baskets which are dipped for a minute into a tank of coloring solution. The dye used is generally 150 Orange I, one of the aniline food colors permitted under the Food, Drug, and Cosmetic act. Other permissible dyes of the same group may be used. The undyed smoked product has a rather pale, unattractive color. The dye solution should be a little weaker than that used for salmon. The dye does not penetrate, is harmless and improves the appearance of the product. Sablefish smoked in the Middle West and Eastern States are not dyed, as the laws in most States in these areas prohibit the artificial coloring of food products.

After dyeing, the pieces of sablefish are placed, skin side down, in shallow trays with wire-mesh bottoms and wooden sides. The tray bottoms are previously coated with an edible oil to prevent the fish from sticking. No two pieces should touch each other. In some plants are stacked on wheeled racks, which are later moved into the smokehouse. In other curing plants the fish are taken directly to fixed racks in the smokehouse. The trays are allowed to drip and drain for a few hours before the smokehouse fire is started, which is usually late in the afternoon. In some instances the sablefish are cured over a low fire, without much smoke, where they are dried and lightly smoked at a temperature around 70°., for 14 to 16 hours. At the end of this period, the fire is built up sufficiently to partially cook the fish, at a temperature around 250° F., for 25 to 30 minutes. Other curers smoke the fish at temperatures varying from 70° to 100° F. for the first 8 to 10 hours. The temperature is then raised to 150° to 170° F. where it is held about two hours for the "cooking."

The barbecued sablefish are then cooled. Some plants use fans to shorten the cooling period. The individual chunks are wrapped in oiled paper, cellophane, or parchment. Sometimes the brand name and other information is printed on the wrapper. In other instances a printed tag is fixed directly to the piece of fish, which is then covered with a plain wrapper. The wrapped pieces are packed in light wooden boxes lined with oiled paper, holding 10 pounds, or in special moisture-proof and oil-resistant heavy fiber boxes containing 10, 20, or 30 pounds of fish.

Barbecued sablefish is a perishable product which spoils after a few days at room temperature, and only quantities that may be readily disposed of are ordered by retail dealers. If the sablefish is to be kept longer than a few days, it should be held in storage at temperatures below freezing. The retailer should hold it in a refrigerated show case at not more than 40° F. Barbecued sablefish is best liked when served steamed or broiled with a parsley butter sauce. As it is somewhat like kippered salmon, it should make good sandwiches and salads.

Sturgeon

Three species of sturgeon are smoked in the United States. The common sturgeon (*Acipenser oxyrhynchus*) is found along the Atlantic coast from Maine to South Carolina but is now taken principally in New Jersey and Virginia. Large-mesh gill nets were formerly used in fishing but most of the catch is now taken in pound nets, incidental to other fisheries. This species has been greatly depleted through overfishing and water pollution.

The green sturgeon (*Acipenser acutirostris*) is distributed from northern California to Alaska, but the commercial catch is now almost entirely taken in the Columbia River area. Small sturgeon are taken in salmon gill nets but the large ones are only caught by the use of set lines specially made of strong rope, and with a large iron hook. The green sturgeon may attain large size. It is reported that the record weight is 1,000 pounds, with a length of 13 feet.

The third species smoked in the United States is known as the Caspian sturgeon (*Huso huso*) imported frozen from the U.S.S.R. Caspian sturgeon may include other species such as *Acipenser mediventris* and *Acipenser ruthenus*, as the common name apparently is used rather loosely for several species. Some fresh and frozen sturgeon is also imported from Canada.

Nearly all the sturgeon used in this country is smoked before going into the markets. Despite the fact that it is a very high priced product, probably the highest on the market, smoked sturgeon is in great demand. The annual production varies between 1,000,000 and 2,000,000 pounds, depending largely on the availability of imported sturgeon. Very little smoked sturgeon is imported. Production at one time was much higher, amounting to 4,000,000 pounds annually, according to Stevenson (1899). Domestic sturgeon was then abundant and the price was much lower, or about 25 cents per pound wholesale, whereas the price is now $1.50 or more.

During the nineteenth century sturgeon for smoking was brine salted and very much like hard-cure salmon and held in barrels until required, when the salt was soaked out in fresh water, as in smoking salmon. This method is no longer used. Fresh fish are used as soon as they reach the plant. Frozen fish, mostly imported, are held in the freezer until required for smoking.

Fresh sturgeon are cut in chunks of about 2 pounds each. The skin is usually left on. These pieces are brined for 10 to 12 hours in a 90° to 95° salinometer brine. On removal from cold storage, frozen sturgeon are thawed in a tank of cold water. The time required until the fish can be handled is about 12 hours. They are then treated exactly like fresh sturgeon. Some smokers roll fish in dry salt, with or without spices crushed in the salt. Others use crushed bay leaves, black peppers, and cardamom seeds. The fish are then packed in a salting butt, with just enough 90° salinometer brine poured in to cover. Some curers prefer a light brine cure, not only 30 minutes. Dry salting is preferred by some curers, who report that it makes the flesh harder and firmer.

After salting, the pieces of sturgeon are rinsed in fresh water to remove excess salt, dirt, slime, or other debris. Sturgeon given a short brining is not rinsed. The chunks of sturgeon are then placed on wire-bottom trays, previously oiled to prevent sticking. No two pieces should touch. The surface moisture on the sturgeon is usually allowed to dry off before the trays are places on racks in the smokehouse. This may be done by a fan. Some smokers still use the old method of stringing the pieces on iron rods, 8 to 10 chunks to the rod, the rod passing through the thin part of the chunk. The rods are then hung in the smokehouse. The trays or rods are fixed from 4 to 6 feet above the smokehouse floor. A low fire is then lighted on the floor and the sturgeon are cured in a light smoke, with all smokehouse ventilators open, for an average of 4 hours. The time is varied according to the weather, and the flavor and texture preferred by the local market. The time must be much longer when the weather is sultry. The minimum time can be used when the weather is breezy and cool. After this period of light smoking, doors and ventilators are closed, a hot fire is built up and the fish is smoked and partly cooked from 1 to 2 hours at a temperature around 175° F.

A few smokers cure the fish for 7 or 8 hours in a rather dense smoke at an even temperature around 125° F. It is reported that 100 pounds of dressed sturgeon will yield 63 to 70 pounds smoked product. Smoked sturgeon are subject to mold if held in chill storage. Some of smoked or kippered sturgeon is canned (Jarvis 1944a).

Whitefish, Lake Herring, and Lake Trout

The smoking of lake herring is centered in the Great Lakes area and the smoked product comprises the major production of the fish smoking industry in the interior region of the United States. The same method is used in smoking lake herring, whitefish, and lake trout. Some lake trout are smoked like salmon. Unscrupulous retailers sometimes attempt to sell lake herring as the more expensive and desirable whitefish, as the finished product is somewhat similar (Tariff Commission 1927). Most of the lake herring and whitefish are imported as frozen fish from Canada. The fish are held in cold storage and are withdrawn as required for smoking.

The frozen fish are first thawed in a tank of cold water. The temperature of the water should be only slightly above the freezing point. Some smokers leave the fish in tanks overnight without changing the water. Others thaw the fish in running water for about four hours. The fish are removed before they are completely thawed as they are firmer and easier to handle. After thawing, the fish are split down the belly to the vent and are eviscerated. The dressed fish are washed thoroughly, taking especial care to remove all traces of blood near the backbone.

After washing, the fish are allowed to drain for a few minutes and are then packed in layers in a brining vat. Some smokers use rectangular vats about 4 feet in width, 10 feet long, and 3 feet deep; others use hogsheads. A small quantity of salt, from 5 to 10 pounds per 100 pounds of fish, is scattered among the fish as they are packed in the vat. The vat is then filled with brine testing 90° salinometer. Sometimes the fish are packed in dry salt, without brine (Jarvis 1945b). Sufficient salt to cover the fish is scattered over each layer. The fish may be taken out of the salt after periods varying from 3 to 10 hours, depending on the flavor desired, the season, and the length of time for which preservation is desired. The salt method is usually followed in summer. Sometimes in winter the fish are cured in brine without the addition of salt. The fish are taken out of brine after a period of from 10 to 16 hours, varying with the preference of the trade, season of year, and amount of fish in the salting vat.

After draining for a few minutes, the fish are hung for smoking. One method is to pass a round smokestick through the right gill and out of the mouth. In some establishments the stick is passed through the eye sockets of the fish; in others an iron wire curved S-shaped is used to attach the fish to the stick, one end of the wire passing through the fish at the head or beneath the nape bone, and the other being hung over the smoke-stick. Some smokehouses use sticks about 2 inches thick and 3 inches wide. Nails are driven through the sticks so that the points project at an angle at each side at 3- or 4-inch intervals. The fish are impaled on these points either under the nape bones or through the vertebrae near the tail.

The method described by Stevenson (1899) is also used:

> "The fish are secured by having stout smokesticks, about 1 1/2 inches thick and 2 1/2 inches wide, in the top of each, and about 1/4 inch from the edge is driven a row of tacks or small wire nails at intervals of about 3 inches, projecting about 1/2 inch above the surface. Ordinary cotton wrapping cord is tied to the wire nail at the end of each stick, the fish being placed with the back of the neck against the stick and the cord passing from one nail around the throat of the fish entering under the gills on each side and then around the next nail, and so on to the end. By having the stick of sufficient width a row of such nails may be placed on each edge, so as to attach a row of fish at each side. This removes nearly all the risk of the fish falling off and their appearance is not marred by holes through which the smokestick has passed."

When the lake herring or whitefish are all hung, the loaded sticks are dipped in fresh water to remove any brine or excess salt from the surface of the fish. The sticks are then hung in a current of air until no moisture is apparent on the surface of the fish. Large mechanical fans are used for this drying operation by some smokers. The sticks are then hung in the smokehouse at 6 to 8 feet from the ground. The smokehouse is of the ordinary type used in smoking other fishery products and has no other special features. Any hardwood may be used for fuel. Hardwood sawdust and chips from woodworking plants are commonly used.

The fish are cured over a light smoke with the smokehouse dampers open for 5 or 6 hours. The temperature at this time should not be more than 90° F., preferably 80° F. At the end of this time the fires are built up, the smokehouse dampers are closed, and the fish are cooked in a dense smoke at 170° to 180° F. for 1 to 2 hours, depending on the height at which the fish are hung, size of the fish, and type of cure demanded in the locality. The fish are allowed to cool, and are then packed in parchment or waxed paper lined wooden boxes for distribution to retailers. The loss of weight in cleaning and smoking averages 50 percent. The product will keep about 10 days.

For sale in markets which prefer the herring well smoked on the inside, the belly cavity is held open by means of toothpicks or somewhat thicker small wooden sticks, two sticks to each fish. This allows the smoke to strike into the belly cavity more completely and gives a product that will keep longer. In general, the western trade prefers the stomach cavity open; the eastern trade does not.

Shrimp

Smoked shrimp is a product which has been prepared commercially only within the past few years. So far as can be determined, it was not sold before 1940. Experimental work in the development of this product was done by the Fish and Wildlife Service, which issued information on two methods, and suggested to fishery interests in the southern states that smoked shrimp had possibilities as a specialty product.

The first method is as follows: Green, headless shrimp are peeled, then washed thoroughly. After draining for a few minutes, the shrimp, in 25 pound lots, are blanched for 5 to 8 minutes in a tank of boiling 50° salinometer brine. If the shrimp are small, the shorter blanching is given; if the weather is humid, the longer time is used. Length of blanching also depends to some extent on market preference. The blanched shrimp are dried, either in the air or by a blower fan, then spread in a thin layer on wire-mesh-bottomed trays previously oiled to prevent sticking. If possible, the meats should not touch each other. The trays are placed on racks in the smokehouse and cured in a dense, cool smoke (temperature not more than 80° F.) for approximately 90 minutes. If a lightly smoked shrimp is desired, 1 hour should be sufficient. The peeled shrimp may be packed in glass tumblers, vacuum sealed, either dry or with vegetable oil. They may also be packaged in glassine or cellophane envelopes holding about 2 ounces.

According to the second method, the shrimp are smoked in the shell (Young 1945). The whole shrimp are first headed, then washed thoroughly in cold running water to remove weeds, dirt, or other extraneous matter. After draining a few minutes the shrimp are placed in a 50° salinometer brine to soak for from 30 to 60 minutes. Length of brining and strength of brine depend on local market preference; some people like salty shrimp, others prefer a mild flavor. The shrimp are then steamed or boiled until they turn pink, indicating that they are sufficiently cooked. The shrimp may also be boiled in a 50° salinometer brine eliminating the brine-soaking step.

The cooked shrimp are spread on wire-mesh-bottomed trays separated so that no two shrimp touch, then dried, either in the open or in the smokehouse, at room temperature (70° to 80° F.) until dry to the touch. A blower or electric fan reduces the drying time and results in a more evenly dried product. Air drying or natural drying may require 2 hours. Artificial drying takes from 10 to 30 minutes.

The shrimp are then cured in a dense cool smoke, preferably at a temperature between 70° and 90° F. Oak or hickory sawdust is preferred for smoking. Softwood sawdust used alone gives the shrimp a bitter taste. A dense smoke is recommended because the smoking time is shorter and the shrimp do not dry out so much. The smoking time varies between 60 and 90 minutes, depending on the size of the shrimp and whether a light or heavy smoke cure is desired. Lightly smoked shrimp in the shell should have a light reddish-brown color. The meat is sweet in flavor with just a tang of smoke. Smoked shrimp in the shell is expected to become a popular cocktail relish. A yield of about 60 pounds of smoked shrimp should be obtained from 100 pounds of green, headless shrimp.

Chapter Twenty-Two

Delicatessen Products

Delicatessen fishery products are classified as such in this handbook because they are largely retailed through food stores of this type. The basic ingredients of these products are salt and smoked fish. Preparation often involves the use of vinegar and spices. While they are not so stable as some types of cured fish, preparation in various ways keeps these products in good condition longer than ordinary fresh fish. More important, however, is the fact that these products are made more appetizing and palatable, thus appealing to consumers who would not relish greatly the ordinary types of cured fish. The recipes given here are only a few of the most popular kinds. There are many others. The products described here have been tested.

Herring Salad

Herring salad is a favorite method of preparing salt herring. There are a great many formulae for this dish. If heat-processed, in glass, a special type of mayonnaise with a much higher solid vegetable fat content is used. The unprocessed herring salad, if packaged and held at a low temperature (about 40° F.), will remain in good condition for several weeks, and is usually more appetizing than the processed type.

This also applies to tuna salad. Sodium benzoate (1/10 of 1 percent) may be used as a preservative, if presence and amount are stated on the label.

Herring Salad, Alaska

10 lb. salt herring	3 lb. apples
2 lb. pickled beans	2 lb. mayonnaise
1 lb. cucumber pickle, sour	1/2 lb. smoked salmon
1 lb. cucumber pickle, sweet	1/4 pound onion
1 lb. mustard pickle	1 bottle (2 oz.) capers

Soak the salt herring in water for about 24 hours, changing the water two or three times, or freshen in running water for 4 to 6 hours. Skin and fillet the herring, removing all bones. Dice into small pieces. Peel and dice the apples, keeping under water until needed, to prevent discoloration. Chop finely the sweet and sour cucumber pickles together with the mustard pickles, pickled beans, and capers. Dice the smoked salmon and grate the onion. Mix all ingredients thoroughly with the mayonnaise. The salad should stand at least 24 hours to blend flavors. This salad is usually served on lettuce with a garnish of hard boiled eggs, capers and nut meats.

Herring Salad, German

25 lb. potatoes, boiled	1 1/2 lb. onions, ground
28 lb. cucumbers, salted	1/2 lb. horse-radish, ground
20 lb. beets, boiled	2 oz. pepper
50 lb. herring, salt	3 lemons, juice and oil (optional).
Paprika to taste	

Freshen herring by soaking in a tank of water from 11 to 24 hours, depending on individual taste. Then skin, fillet, and dice fish. Dice cucumbers, beets, and potatoes and mix with herring. Mix horseradish, onions, and pepper with mayonnaise. The juice and oil of 3 lemons may be added, if desired. Fold mayonnaise mixture lightly but thoroughly into the other ingredients. Package and store at 34° to 40° F.

Herring Salad, Italian

10 lb. boiled tongue
6 lb. apples
20 lb. boiled potatoes
1 1/2 lb. ground onions
1/2 lb. horse-radish
30 lb. salt cucumber
20 lb. salt herring
6 lb. mayonnaise
10 lb. carrots

Curry powder, mace, and pepper to taste. Juice and oil of 3 lemons optional.

Freshen the salt herring by soaking in water overnight. Skin and fillet, removing all bones. Dice. Chop celery fine. Peel apples, potatoes, and carrots and cut into small dice. Dice tongue and cucumbers. Mix all these ingredients thoroughly. Blend onions, horseradish, and spices with the mayonnaise. The juice and oil of 3 lemons may also be added, if desired. Then add mayonnaise to other ingredients. Store in a cool place. Though this herring salad is known as Italian it was obtained from a German source.

Herring Salad, Swedish

5 lb. salt herring
3 lb boiled veal
1/2 lb. ham
1 pint white vinegar
2 lb. apples
1/2 lb. onions
1/2 lb. sour pickles
2 lb. French dressing
2 lb. boiled potatoes
2 lb. boiled beets
1/2 lb. chopped anchovy fillets

Soak herring in fresh water overnight (12 hours). Skin and fillet, removing all bones. Wash and drain, then cut into small cubes. Mix with chopped veal, ham, anchovies, onions, diced potatoes, beets, apples, and pickles. Add French dressing and vinegar. Mayonnaise may be substituted for French dressing, or the French dressing may be diluted with mayonnaise. Stir thoroughly together and store in a cool place for 48 hours. Arrange on platter and garnish with sliced hard boiled eggs, capers, small pickled onions, parsley, and mayonnaise.

Smoked Herring Salad

8 oz. canned smoked herring fillets in oil

4 oz. anchovy fillets	1 1/2 cups whipping cream
2 lb. boiled potatoes	2 lb. boiled beets
2 lb. tart cooking apples	1 cup finely chopped onion
1/4 cup vinegar	4 Tbs. sugar

Cut herring and anchovy fillets into very small cubes. Peel boiled beets and potatoes, then cut into larger cubes. Peel apples and cut into the smallest possible cubes. Apples are intended to give freshness to the taste, but should not be visible. Mix fish, beets, potatoes, apples, and onion. Add vinegar in which sugar has been dissolved. Ingredients should be mixed very lightly so that the cubes are not broken. Finally fold in the cream whipped to the consistency of a thick sauce. It must not be too thick or it will curdle when mixed. Pack in containers, to be sealed and held at a temperature of 40° F. until retailed.

Tuna Salad I

No two commercial tuna salads are prepared by exactly the same formula, but they do not show the wide variety characteristic of herring salad. The recipe given here is typical. It is offered, however, only as a guide. The same recipe with minor variations to suit the taste may be used for salmon salad.

5 lb. tuna	2 1/2 pt. finely diced celery
1 pt. finely diced sweet pickle	12 diced hard boiled eggs
	1 1/2 pt. mayonnaise
5 lemons	1 1/2 tsp. salt
3/4 tsp. pepper	1/2 lb. hydrogenated vegetable shortening

Flake tuna, dice celery, pickles, and eggs. Extract juice of lemons. Combine all ingredients except lemon juice. Season with salt and pepper, and sprinkle lemon juice over the whole. Combine shortening with mayonnaise, then mix with other ingredients. In some formulae one pint of diced cucumber is included. Others use sour instead of sweet pickles. Still others use capers in place of pickles. These variations are entirely a matter of taste. Package and hold at 40° F. until retailed.

Tuna Salad II

This is the only commercial recipe for tuna salad that differs widely from the one given above.

5 lb. tuna	5 lb. diced boiled potatoes
2 1/2 cups diced dill pickles	7 1/2 cups French dressing

Salt and cayenne to taste.

Flake tuna, mix with other ingredients and season. Fill the salad into containers and hold at 40° F. until retailed.

Salmon Salad

5 lb. salmon	5 lb. diced boiled potatoes
2 cups diced onion	8 cups sour-cream dressing

Salt, pepper, and fresh dill to taste.

Combine ingredients, then season with salt, pepper, and dill, mixed. A cup of diced dill pickle is sometimes used in place of fresh dill. sometimes a half-cup of fresh grated horseradish is added to the sour-cream dressing, omitting fresh dill or dill pickles. To make this amount of sour cream dressing take:

6 cups sour cream	6 eggs beaten lightly
1 cup vinegar	4 Tbs. salt
2 Tbs. mustard	3/4 tsp. pepper

Add eggs, vinegar, and dry ingredients to cream, mixing thoroughly. Cook dressing in a double boiler stirring constantly, until the mixture begins to thicken. Package and store like herring or tuna salad.

Fish Cakes

Fried fish cakes are sold rather widely in delicatessens and at prepared food counters of department stores in the Atlantic coastal area. This product has possibilities for other sections of the country.

10 lb. potatoes	5 lb. shredded salt cod
8 lb. eggs, beaten	8 oz. butter or margarine
1 lb. diced onions	1/4 oz. pepper

Soak fish in cold water for several hours. Then shred or break into small pieces. Place the fish in a kettle and add water. Heat to boiling point, and drain off water. Boil potatoes and fish together until done. Drain off liquid and put fish and potatoes in meat grinder with 1/4-inch plate. Put ground mass in an electric mixer, add beaten egg, margarine, onion, and pepper and beat until ingredients are thoroughly mixed. Onions are not used unless the trade has a preference for them. Form mixture into cakes of the usual size, dip in beaten egg, roll in fine bread crumbs, and fry until a light brown. The cakes are not packaged but sold from refrigerated display cases held at approximately 40° F.

Crab Cakes

Crab cakes are sold rather extensively in delicatessens and food stores in Maryland, Virginia, and North Carolina, and also to some extent in other middle Atlantic and southern states. The product varies widely in quality, usually according to the quality of ingredients, cooking fat, and care in preparation. Any good standard recipe is satisfactory. The following formula is recommended:

5 lb. crab meat	2 1/2 cups soft bread crumbs
2/3 cup minced onion	6 oz. butter
5 beaten eggs	2 Tbs. dry mustard
2 Tbs. Worcestershire sauce	Salt, paprika to taste

Melt butter, add onion, and cook until soft and yellow. Add other ingredients and mix together. If the texture is too dry add small amount of milk. Form mixture into cakes of the desired size (usually about the same size as cod fish cakes), dip in beaten egg, roll in bread crumbs, and fry in deep fat (about 375° F.) until brown. Drain cakes on fat-absorbent paper. Handled like fish cakes.

Seviche

Seviche is typically a South American preparation. It is sold commercially in the United States principally in the Latin-American quarters of our large cities. It is essentially cubed raw fish, preserved by marinating in sour-orange juice. This is not the juice of unripe oranges but a special variety of fruit with a very sour juice, which is grown in southern Florida. Lime juice is generally used as a substitute in the United States but in South America it is considered inferior to sour-orange juice. Lemon juice is never used. The fish should be fresh and firm fleshed. Corvina is most favored in South America. Striped bass has been found to be the best substitute in North America, though sea trout (weakfish) may also be used. The ingredients are:

10 lb. fish	5 large onions
3 cloves of garlic	5 cups lime juice
12 to 15 yellow chili peppers	Salt and cayenne to taste

Scale fish and clean thoroughly. Fillet fish, removing backbone and cut fillets into half-inch cubes. Wash and drain diced fish. Slice onions thin, then mix with garlic, ground fine. Mix fish and onion-garlic mixture together in a large bowl and season with salt to taste. Slice peppers and add. The yellow hot peppers of Louisiana, preserved in glass, are acceptable for the purpose. Stir a little cayenne pepper into lime juice and pour over the whole. Tartaric acid, dissolved in water, is sometimes used instead of lime juice. The seviche is allowed to stand overnight before use. This preparation will remain in good condition about 5 days at average room temperatures, and longer under refrigeration.

Seviche is also made from shellfish such as clams or mussels and shrimp or fresh-water crayfish. The general method of preparation is as described above. The meat of the crustacea is used raw of "green." The clams and mussels are shucked raw. Only the "beard" is removed from the mussels. The dark body mass or "stomach" is separated from the clams.

Gravlax (Marinated Salmon)

10 lb. salmon	1 pt. olive oil
1 cup salt	1 cup brown sugar
1/4 cup fresh, chopped dill	1 Tbs. allspice
1 Tbs. white pepper	1 Tbs. saltpeter
1 Tbs. mace	

This is a favorite Swedish delicatessen dish. Clean and wash the salmon thoroughly. Then split lengthwise, removing the backbone. Rinse fillets, drain and wipe dry. Rub oil well into both sides of the fish halves. Mix together with salt, sugar, saltpeter, white pepper, ground allspice, and mace. Coat both sides well with mixture, rubbing into flesh. Place chopped dill on top. Place both halves together and tie tightly with string. Set in a shallow pan and press under moderate weight for 48 hours. Serve in thin slices. May be packaged in glass. Store at 34° F.

Lutefisk

Lutefisk is a food product of Scandinavian origin, prepared from stockfish. Lutefisk is prepared by a special process. Stockfish is split, dried, unsalted codfish (see section on curing stockfish). Hard-dried or export-cure salt cod is sometimes used in making lutefisk, but is considered inferior to stockfish as a raw material.

Lutefisk is sold at most of the better groceries and delicatessens in the middle west and at some delicatessen shops in the eastern United States. The large importers and dealers in fancy food stuffs in Chicago and New York City always have an abundance of lutefisk at the Christmas season, for it is a favorite dish at Christmas and New Year among all Scandinavian people.

To prepare lutefisk for this trade, proceed as follows: Cover the stockfish with clear cold water and allow it to soak for 4 days, being sure to change the water every day. The fish should be kept in a cool place preferably out of doors as the odor is objectionable to some people, but care should be taken to prevent freezing as this ruins the product. When the fish has been sufficiently soaked, cover with cold water to which has been added a solution of soda and lime; for 5 pounds of fish use 2 cups of washing soda and 3 or 4 cups of slack lime to about 12 gallons of water. A large container such as a salting butt is used for soaking. Let the fish soak in this "lute" for 3 or 4 days. Pour off the solution and again soak the fish in clear cold water for at least 3 days to remove all of the chemicals, changing the water every day. Keep in a cool place or the fish will spoil. The lutefisk should now have the desired consistency and flavor.

If directions for cooking are requested the standard recipe is: Cover fish with cold water and allow it to come to a boil, adding a little salt. Boil about 10 minutes, then pour off water. The cooked lutefisk should have a white and flaky appearance, breaking into pieces as it is placed on a platter. Serve with melted butter or a white sauce, together with boiled or mashed potatoes, dashed with pepper and salt to taste.

Literature Cited

Ackerman, Edward A., 1941. *New England's Fishing Industry.* 303 pp., 122 fig. Chicago.

Anderson, A.W. and Power, E.A., 1946. Fishery Statistics of the United States, 1942. *Statistical Digest 11,* U.S. Fish and Wildlife Service. 248 pp. Washington.

Anderson, Clarence L., 1925. *Preserved Pickled Herring.* University of Washington Pub. in Fish., 1(1):1-64.

_____and Robert K. Pederson, 1947. *The Smoke-curing of fish and the application of a controlled method to the process.* Tech. Report No. 1, Department of Fisheries, State of Washington. 20 pp, 3 fig. Olympia, Wash.

Anonymous. (Undated.) *The History of Salt.* Morton Salt Co. 32 pp., 20 fig. Chicago.

Anonymous, 1944. *That Pinch of Salt.* General Foods Corp. 12 pp., 7 fig. New York.

Anonymous, 1945. *Fishery Resources of the United States.* IV-135 pp. Senate Doc. No. 51, 79th Congress, First Session. Washington.

Anonymous, 1946a. Moderno procedimiento para salazon de pescado. *El Agricultor Venezolano.* Vol. X:No. 113. Feb.-Mar. 1946. Caracas.

Anonymous, 1946b. *Notes on the Torry Research Station Fish Smoking Kiln.* Ministry of Food, Great Britain. 5 pp., mimeo. Aberdeen.

Arnason, Svein, 1941. *Observations on the production of salt fish.* 1-91 pp. Reykjavik.

Beatty, S.A., 1942. Putty Fish. Fisheries Research Board of Canada. *Progress Reports of the Atlantic Coast Stations,* No.31. Feb. 1942. 3-5 pp.

Bitting, A.W., 1911. *Preparation of the cod and other salt fish for the market including a bacteriological study of the causes of reddening.* U.S. Dept. of Agr., Bur. Chem. Bul. 133. 54 pp. Washington.

Borodin, N.S., 1920a. The Preparation of Caviar. *Fishing Gazette.* Vol. 37, No. 9. Sept. 1920. 10-11 pp. (Also U.S. Bur. Fish. Mem. S169) 2 pp. New York.

_____1920b. *Methods of Preserving and Curing Fish Products* (Russian). 54 pp., 12 fig. New York.

Boury, Maurice, 1932. Recherches sur la morue salée. *Revue de l'Office des Pêches Maritimes.* t. V, fasc. 2, 193-222 pp. Paris.

_____1934. Etudes sur le salage du poisson. *Revue de l'office des Pêches Maritimes.* t. VII, fasc. 2, 193-222 pp. Paris.

Burton, L.V., 1943. Infrared Speeds Fish Drying. *Food Industries,* Vol. 15, No. 12. 80-81 pp., 4 illus. New York.

Canadian Fisherman, 1942. Canadian Fisheries Manual, 1942: *The Story of Canada's Fishing Industry.* 116 pp., illus. Gardendale, Que.

Canadian Legion, 1945. Vocational courses for service personnel. *Fishing textbook No. 2,* 116 pp., 34 fig. Ottawa.

Caribbean Commission, Anglo-American, 1945. *Guide to commercial shark fishing in the Caribbean area.* U.S. Dept. Interior, Fish and Wildlife Service. Fish. Leaflet 135. 149 pp., 56 fig. Washington

Carter, Neal M., 1932. *The salting of herring.* Biological Board of Canada. Progress Reports No. 13. 3-7 pp. July, 1932.

Census, U.S. Bureau of, 1911. *Fisheries of the United States,* 1908. Special Reports Bur. Census. Dept. Comm. and Labor. 324 pp. Washington.

Classen, T.E.A., 1946. *The Tuna Industry of Southern Spain.* U.S. Dept. Interior, Fish and Wildlife Service. Fish Leaflet 188. 28 pp. Washington.

Cobb, John N., 1919. *Canning of Fishery Products.* 217 pp., illus. Seattle.

_____1926. *Pacific Cod Fisheries.* U.S. Bur. Fish Doc. No. 1014, 385-499 pp., 7 illus., revised ed. Washington.

_____1930. *Pacific Salmon Fisheries.* U.S. Bur. Fish., Doc. No. 1092, 409-704 pp., 48 illus., revised ed. Washington.

Coceani, Bruno, 1940. *La Pesca Italiana.* Edizioni Usila. 141 pp., illus. Rome.

Cooper, D. LeB., 1937a. A satisfactory small smokehouse. Fisheries Research Board of Canada. *Progress reports of the Atlantic Coast Stations,* No. 19, May 1937, 7-9 pp.

_____1937b. Fish drying. Fisheries Research Board of Canada. Prog. reports of the Atlantic Coast Stations, No. 20, Aug. 1937, 8-10 pp.

_____1938. Studies on salt fish. III. Equilibrium constant of salt fish muscle. *J. Fish Res. Bd. Canada.* Vol.4, No.2, Aug. 1938. 136-140 pp. Toronto.

Cooper, D. LeB. and Linton, E.P., 1934. The smoke curing of fresh fillets. Part I. Biological Board of Canada. *Progress Reports, Fish Exp. Station (Atlantic)* No. 12, Oct. 1934, 7-9 pp.

_____1936. The preparation of fillets of fish for smoking. *Jour. Biol. Bd. Canada,* Vol. 3, No.1. November 1936, 1-11 pp. Toronto.

Cutting, C.L., 1942. Engineering problems of the smoke curing of fish. *(Chem. and Ind.*, 1942, 61, 365).

_____1944. The effect of refrigeration and dehydration on the composition, keeping quality, and culinary properties of fish. (Proc. Nutr. *Soc. 1944,2, 113*) F.I.B. Mem. No. 464.

_____1945. Some of the principles involved in the smoke curing of fish. *(Chem. and Ind.*, 1945, Mar.9 (No. 9) Vol. 66, F.I.B. Mem. No. 465.

Cutting, C.L., and Reay, G.A., 1944. The dehydration of fish. Part I, General, (*Chem. and Ind.* 1944, Feb 5, 47 pp. F.I.B. Mem. No. 434.

Dieuzeide, R., and Mercier, A., 1941. *Le fumage du poisson*. Gouv. Générale de l'Algerie, Doc. et renseign. Agricole. Bull. 44: 28pp., 8 fig. Algiers.

Dieuzeide, R., and Novella, M., 1942. *Essai sur la technique des salaisons de poisson*. Gouv. Générale de l'Algerie, Algiers.

Duthie, R.J. ("Viking"), 1911. *The art of fish curing.* 111 pp., illus., Rosemont Press, Aberdeen.

Dyer, W. J., 1941. Salting by weight. Fisheries Research Board Canada. *Progress Reports of the Atlantic Coast Stations,* No. 30, Nov. 1941., 8-9 pp.

_____1942a. The salting of codfish. I. Fisheries Research Board Canada. *Progress Reports of the Atlantic Coast Stations.* No. 32, July 1942. 6-8 pp.

_____1942b. The salting of codfish. II. Fisheries Research Board Canada. *Progress Reports of the Atlantic Coast Stations.* No. 33, Nov. 1942, 3-6 pp.

Fiedler, R.H., 1943. *Fishery statistics of the United States*, 1940. Stat. Digest No.4, U.S. Fish and Wildlife Service, 174 pp. Washington.

_____1945. Fishery statistics of the United States, 1941. *Stat. Digest No. 7, U.S. Fish and Wildlife Service.* 174 pp. Washington.

Fiedler, R.H., Jarvis, N.I., and Lobell, M.J., 1943. *La Pesca y las industrias pesqueras en el Peru.* Compañia Administradora del Guano. 371 pp. Lima.

Fillon, Robert, 1929. Recherche des meilleurs seis pour le salage de la morue. *Revue de l'Office des. Péches Maritimes*, t.II, fasc.3. 295-304 pp.

Finn, D.B., 1936. Preservation by salting. *Biological Board Canada. Progress* Reports. Fish Exp. Station (Atlantic) No. 18 Nov. 1936. 7-11 pp.

Foote, W.H., 1938. *Methods of production and curing of Norwegian fish.* Economic Bull. No. 6, Dept. Nat. Resources. 1- 12 pp. St. Johns.

Frank, M., and Hess, Ernest, 1941a. Studies on salt fish. IV. Survival of Eberthella typhosa and Escherichia coli on salt fish. Vol. 5, No. 3, 249-252 pp. *Jour. Fish. Res. Bd. Canada.*

_____1941b. Studies on salt fish. V. Studies on Sporendonema epizoum from "dun" salt fish. *Jour. Fish. Res. Bd. Canada.* Vol. V, No.3, 276-286 pp.

_____1941c. Studies on salt fish. VI. Halophilic brown molds of the genus Sporendonema emend. Cifferi et Raedelli. *Jour. Fish. Res. Bd. Canada.* Vol. V, No. 3, 287-292 pp.

Gibbons, N.E., 1935. The control of reddening of salt fish. *Biol. Bd. Canada.* Progress Reports, Fish. Exp. Station (Atlantic) No. 14, June 1935, 13-14 pp.

_____1937. Studies on salt fish. I. Bacteria associated with the reddening of salt fish. *Jour. Biol. Bd. Canada.* Vol. 3, No.1, 70-76 pp. Toronto.

Goode, G. Browne, 1884. *Fisheries and fishery industries of the United States.* Five Vols., 1884 to 1887 plus 2 Vol. plates total 277 plates. U.S. Fish Commission, Washington.

Griffiths, Francis P., and Lemon, J.M., 1934. *Studies on the smoking of Haddock.* U.S. Bur. Fish., Inv. Rpt. No. 20, 12 pp. Washington.

Gruvel, A., 1922. En Norvege. *L'industrie des pêches.* Office Scientifique et Technique des Pêches Maritimes. Notes et Mémoires, No. 16, 170 pp., 46-33 fig. Paris.

Hampton, William F., 1938. *Pink fish, its causes and prevention.* Serice bull. No. 4, Dept. Nat. Resources. 1-11 pp. St John's.

Hess, Ernest, 1929. The bactericidal action of smoke. *Contrib. to Canadian Biol. N.S.* Vol.1, No. 4, 29.76 pp.

_____1940a. Instructions for sterilizing fish curing plants, etc. Fish Research Board Canada. *Progress Reports Atlantic Coast Stations.* No. 27, April 1940, 3-5 pp.

1940b. "Dun" salt fish. Fisheries Research Board of Canada. *Progress Reports of the Atlantic Coast Stations,* No. 28, Dec. 1940, 17-99 pp.

_____1942. Studies on salt fish. VII. Red halophilic bacteria in seawater and fish slime and intestines. *Jour. Fish. Res. Bd. Canada.* Vol. 5, No. 5, 438-439 pp.

_____1942a. Studies on salt fish. VIII. Effects of various salts on preservation. *Jour. Fish. Res. Bd. Canada.* Vol. 6, No. 1, Dec. 1942, 1-9 pp. Toronto.

_____1942b. Studies on salt fish. IX. Effects of environment upon growth of red halophilic bacteria. *Jour. Fish. Res. Bd. Canada.* Vol. 6, No. 1, Dec, 1942, 10-16 pp. Toronto.

Hess, Ernest and N.E. Gibbons, 1942. Studies on salt fish. X. Effects of disinfectants and preservatives on red halophilic bacteria. *Jour. Fish. Res. Bd. Canada.* Vol. 6, No.1, Dec. 1942, 17-23 pp.

Hoffman, Herbert, Undated. *Lehrbuch der Fischverwertung fur die Fischindustrie.* (Eighth revised ed., Lehrbuch der Fischbereitung. Wilhelm Dunker). 197 pp., illus. Stettin.

Holdsworth, E.W.H., and Young, Archibald, 1877. *Sea and salmon fisheries.* 300 pp., illus. London.

Innes, Harold A., 1940. *The cod fisheries.* XVII plus 520 pp. New Haven.

Jacobsen, Edard, 1926. Handbuch f129r die konserven industrie, *konserven fabriken und den konserven grossbetreib.* Vol. II, 1459 pp., 635 fig. Berlin.

Jarvis, Norman D., 1932a. *The fisheries of Puerto Rico.* U.S. Bur. Fish. Inv. Rept. No. 13, 41 pp., 8 fig. Washington.

_____1932b. *Stockfish.* U.S. Bur. Fish. Mem. S-327, 3 pp. (mimeo.). Washington.

_____1933. Smoking fish in southern States. *U.S. Bur. Fish., Mem.* S-331, 4 pp. (mimeo.). Washington.

_____1935. Utilization of salmon eggs. *U.S. Bur. Fish. Spec. Memo.* 1487-a, 3 pp. (mimeo.). Washington.

_____1936a Alaskan Scotch-cured herring. *U.S. Bur. Fish. Mem.* S-342, 7 pp. (mimeo.). Wahington.

_____1936b. Mild curing, pickling, dry-salting and smoking salmon. *U.S. Bur. Fish. Mem.* S-343, 13 pp. (mimeo.). Washington.

_____1943a. *Home preservation of fishery products.* Dept. Interior, Fish and Wildlife Service. Fish leaflet No. 18, 25 pp. (Mimeo.). Washington.

_____1943b. *Brine salted herring.* U.S. Dept. Interior, Fish and Wildlife Service, Fish. Leaflet No. 38, 3 pp.(mimeo.). Washington.

_____1944a. *Principles and methods in the canning of fishery products.* Research Rept. 7, U.S. Fish and Wildlife Service, 1-366 pp., 66 fig. Washington.

_____1944b. *Brine-cured cod.* U.S. Dept. Interior, Fish and Wildlife Service. Fish. Leaflet No. 44, 2 pp. (Mimeo.) Washington.

_____1945a. *Brine salted mullet.* U.S. Dept. Interior, Fish and Wildlife Service. Fish. Leaflet No. 51 (Mimeo.) 4pp. Washington.

_____1945b. *Smoking lake herring, whitefish, lake trout and carp.* U.S. Dept. Interior, Fish and Wildlife Service. Fish. Leaflet No. 122 (Mimeo.) 6 pp. Washington.

_____1945c. *Dry-salting mullet, red drum (channel bass) and kingfish (king mackerel).* U.S. Dept. Interior, Fish and Wildlife Service. Fish. Leaflet No. 136 (Mimeo.), 7 pp. Washington.

Jarvis, Normen D., and R.W. Hines, 1943. *Salted river herring (alewives) in Chesapeake Bay.* U.S. Dept. Interior, Fish and Wildlife Service, Fish. Leaflet No. 26, 4 pp. (Mimeo.). Washington.

Kasatkin, F.S., 1944. *Technology of fishery products* (Russian). 655 pp., illus. Moscow.

Klie, A.H.D., 1920. *Scotch method of curing herring.* U.S. Bur. Fish., Mem. S-58 pp. (Mimeo.). Washington.

Labrie, A., and N.E. Gibbons, 1937. Studies on salt fish. II. The effect of salt concentration on preservation. *J. Biol. Bd. Canada.* Vol.3, No.5, 439-449 pp.

LeGall, Jean, Undated. La peche en Norvege. Memoires, *Serie Speciale, Office Scientifique et Technique des Peches Maritimes,* No. 4, 87 pp., 58 figs. Paris.

Leim, A.H., and A.S. McFarlane, 1932. The smoking of fillets and haddies. Biological Board Canada. *Progess Reports. Fish Exp. Station (Atlantic).* No. 4, July 1932, 6-8 pp.

Lemon, James M., 1932. Controlled Smoking. *Fishing Gazette,* Vol. 49, No. 8, 12-14 pp. New York.

Linton, E.P., and H.V. French, 1945. Factors affecting deposition of smoke constituents on fish. *Jour. Fish. Res. Bd. Canada,* Vol. VI, No. 4, 338-348 pp. Toronto.

Linton, E.P., and A.L. Wood, 1943. A tunnel smoke-house for the smoke curing of fish. Fisheries Research Board Canada. *Progress Reports of Atlantic Coast Stations.* No. 34, Nov. 1943, 10-14 pp.

_____1944. Salt fish dryer. Fisheries Research Board Canada. *Progress Reports of the Atlantic Coast Stations.* No. 35, March 1944, 26-28 pp.

_____1945. Drying of heavily salted fish. *Jour. Fish. Res. Bd. Canada.* Vol. VI, No. 5, 380-391 pp.

Looker, C.D., 1940. *Salt, the aristocrat of minerals.* International Salt Co., Inc. 20 pp. Scranton.

Makarowa, T.I., 1939. The qualitative composition of smoke liquid. *Trans. Ins. Marine Fish. and Ocean. of the U.S.S.R.* Vol. IX, 34-38 pp. Moscow.

McPherson, Norman L., 1935. *The dried codfish industry.* Service Bull. No. 1, Dept. Natural Resources, 1-56 pp. St. Johns.

Morison, Samuel Eliot, 1941. *The maritime history of Massachusetts.* (2d ed.) 421 pp., illus. Boston.

Myles, H.W., G.A. Reay, and H.E.M. Farrer, 1946. Smoke curing of fish in Germany. *Fish Trades Gazette:* Vol. LXIV, No. 3302, Sept 7, 1946. 6-7 pp. London.

Nelson, Thomas, 1934. How to smoke fresh-water fish. *Fish. Gazette,* Vol. 51, No. 12, November 1934, 14 pp. New York.

Otero, Enrique, and M. Jesus del Val, 1943. Influencia del contenido graso de las especies en la salazon del pescado. ns. *Esp. de Oceanografica. Notas y resumes,* Serie II, Numero 113, 16 pp. Madrid.

Petterson, Olive Gair, 1918. The smoking of "haddocks" for Canadian markets. *Contrib. to Canadian Biol. 1917-18,* No. X, 175-78 pp. Ottawa.

Radcliffe, Lewis, 1925. *Caviar: What it is and how to prepare it.* U.S. Bur. Fish. Ec. Cir., No. 20, 8 pp., 3 fig. Washington.

Radcliffe, William, 1921. *Fishing from the earliest times.* IX plus 478 pp., 56 fig. London.

Reay, G.A., 1936. The salt curing of herring. *(Jour. Soc. Chem. Ind.* 1936, 55, 309 T.) Food Inv. Bd. Mem. No. 314. London.

Rector, Thomas M., 1925. *Scientific preservation of food.* 217 pp., illus. New York.

Scofield, W.L., 1925. *Mild curing of salmon in California.* U.S. Bur. Fish. Doc. No. 983, 1-14 pp. Washington.

Scott, Geo. C., and Harold G. Wolff, 1936. *Dehydration of Fish.* U.S. Bur. Fish. Mem. S-181, 5 pp. (Mimeo.). Washington.

Seale, Alvin, 1914. Preservation of commercial fish and fishery products in the tropics. *Philippine Jour. Sci.* Vol. 9, No. 1, Feb. 1914. 1-18 pp., 4 fig. Manila.

_____1917. Sea products of Mindanao and Sulu III: Sponges, tortoise shell, corals, and trepang. *Philippine Jour. Sci.* Vol. 12, No. 4, July 1917. 191-212 pp., 4 fig. Manila.

Shewan, J.M., 1944. The effect of smoke curing and salt curing on the composition, keeping quality, and culinary properties of fish. *(Proc. Nutr. Soc.* 1944, 2, 105). F.I.B. Mem. No. 463.

_____1945. Some of the principles involved in the smoke curing of fish. (*Chem. and Ind.*, 1945, Mar. 31. (No. 13): 98) F.I.B. Mem. No. 468.

Sigurdsson, A.J., 1947. Salting herring in Iceland. *Food Ind.,* Vol. 19, No. 4, 73-76 pp. New York.

Smidth, J.K., 1873. Historical observations on the condition of the fisheries among the ancient Greeks and Romans, and on their mode of salting fish. *Report, Comm. Fish and Fisheries*, Part III, App. A, 1-20 pp., 1876.

Stahmer, Max., 1925. *Fischhandel und fischindustrie.* 431 pp., 36 fig. (Rev. ed.). Stuttgart.

Stansby, Maurice E., 1944. *Dehydration of fishery products.* U.S. Dept. Interior, Fish and Wildlife Service. Fish Leaflet 120. 4 pp. (Mimeo.).

Stansby, Maurice E., and Francis P. Griffiths, 1943. *Preparation and keeping qualities of lightly smoked mackerel.* Res. Rpt. 6, U.S. Fish and Wildlife Service. 1-14 pp. Washington.

Stevenson, Charles H., 1899. The preservation of fishery products for food. *U.S. Fish. Comm. Bull.* 1898. 335-563 pp., plates I-XLIII. Washington.

Surzhin, S. N., 1937. Liquid for the wet smoking of fish. *Trans. Inst. Mar. Fish. and Ocean of the U.S.S.R.,* Vol.6, 71-102 pp., 6 fig. Moscow.

Tanner, Fred Wilbur, 1944. *The microbiology of foods.* 1196 pp. (second ed.). Champaign.

Tariff Commission, U.S., 1927. Lake fish. *Tariff Inf. Series No. 36,* 175 pp., illus. Washington.

_____1933. *Fishery products. Report No. 69,* second series, 351 pp., illus. Washington.

_____1943. *Salted fish.* 89 pp. Mimeo. Washington.

_____1944. *Treaties affecting the northeastern fisheries.* Rept. No. 152, second series, XXIV plus 167 pp. Washington.

Tarr, H.L.A., 1945. Water micro-organisms and volatile bases in dehyrdated fish. *Jour. Fish. Res. Bd. Canada.* Vol. 6, No.4, May 1945. 303-310 pp. Toronto.

Taylor, Harden F., 1921. *Improvements in process of salting river herring, especially adapted to warm climates.* U.S. Bur. Fish. Doc. No. 903, 7 pp. Washington.

_____1922. *Principles involved in the preservation of fish by salt.*

U.S. Bur. Fish. Doc. No. 919, 22 pp. Washington.

Tressler, D.K., 1920. *Some considerations concerning the salting of fish.* U.S. Bur. Fish. Doc. No. 884, 54 pp. Washington.

_____1923. *Marine products of commerce.* 762 pp., 257 fig. New York.

Uriarte, Luis Bellon, 1926. *Industria del atun en España.* Ins. Esp. de Ocean, No. 3, 98 pp. 30 fig. Madrid.

Victorin, Heinrich, 1906. *Die meeresprodukte.* 455 pp., illus. Vienna.

Watson, D.W., and Ernest Hess, 1936. The preparation of "marinated" herring. Biological Board of Canada. *Progress reports. Fish Exp. Station (Atlantic)* No. 18, 3-6 pp. November 1936.

Wilson, Anna M., 1938. *Production and composition of commercial salt.* Service Bull. No. &, Dept. Nat. Resources. 14 pp. St. John's.

Young, Leo, 1945. Smoking Shrimp. *Fish. Mkt. News,* Vol. 7, No. 9, 2-4 pp. Washington.

Young, Leo, and C.F. Lee., 1943. Dehydration of fishery products. *Fish. Gazette,* Vol. 60, No. 13, p. 64.

Curing of Fishery Products

Index